Study on Safety and Adverse Effects of Ivabradine in Acute Myocardial Infarction

First Printing: 2020

ISBN: 978-1-79486-021-6

Affiliation of Author

Shirisha Marupakala,
Guru Nanak Institutions Technical Campus-School of Pharmacy,
Ibarhimpatnam, Ranga Reddy District, Hyderabad,
Telangana, India- 501506.

Keerthi Thanneru
Guru Nanak Institutions Technical Campus-School of Pharmacy,
Ibarhimpatnam, Ranga Reddy District, Hyderabad,
Telangana, India- 501506.

Murali Krishna Chirumamilla,
Guru Nanak Institutions Technical Campus-School of Pharmacy,
Ibarhimpatnam, Ranga Reddy District, Hyderabad,
Telangana, India- 501506.

Lakshmi Thakkalapally,
Guru Nanak Institutions Technical Campus-School of Pharmacy,
Ibarhimpatnam, Ranga Reddy District, Hyderabad,
Telangana, India- 501506.

www.lulu.com
Lulu Press, Inc
627 Davis Drive, Suite 300, Morrisville, NC 27560.

Study on Safety and Adverse Effects of Ivabradine in Acute Myocardial Infarction

Author

Shirisha Marupakala
Keerthi Thanneru
Murali Krishna Chirumamilla
Lakshmi Thakkalapally

Editor

Sagar Pamu

2019

About Author

Shirisha Marupakala, pursued Pharm. D at Guru Nanak Institutions Technical Campus-School of Pharmacy (GNITC-SOP). She attended and presented in various conferences and seminars on different topics related to clinical pharmacy. She has an experience of patient counselling during medical camps and she achieved top score during academics.

Keerthi Thaneeru, pursued Pharm. D at GNITC-SOP. She has skills in ADR detection & prescription analysis. She also attended and presented in various conferences & seminars.

Murali Krishna Chirumamilla, pursued Pharm.D at GNITC-SOP. He has an experience in patient counselling during medical camps. He also attended and presented in various conferences & seminars.

Lakshmi Thakkalapally, Ph. D, Professor at GNITC-SOP. She authored for 20 papers and 2 book publications. She also attended and presented in various conferences and seminars. She played a role of convener for ICIPS-2018 & ICIPS-2020.

Acknowledgement

At the very outset, we thank God, the Almighty for showering his blessings and being a source of guidance and wisdom throughout the study without which no human achievement is possible.

We are indebted to our beloved **Parents** without whose encouragement and help our professional career would never seen the light of the day.

As we walk along the path of life, we have the pleasure of meeting people who search our life in such a way that it never is the same gain…it may be small thoughtful things they do, a smile, a helping hand, a word of encouragement or just by mere presence they make our life worth living.

Accomplishing this project has been a great learning and very fulfilling experience. There have been many people who have come along side and helped in conceiving, designing, and executing this project. I would like to place a record and my sincere appreciation for their contribution.

"Words cannot be said nor written for obligation and indebtedness"

We take immense pleasure in thanking my research guide **Dr. Praneeth Polamuri, MD, DM, Department of cardiology, Care Hospitals** for his rejuvenating inspiration, kind co-operation, valuable guidance, suggestions and encouragement throughout the progress of this work which helped me to complete every aspect of this project work and for his generous offer to use the available data on Primary PCI patients.

We wish to extend my sincere thanks to my co-guides **Dr. Praveen, MD, DM, Dr. Rahul Agarwal, Intensive Cardiologist,** for their kind co-operation, valuable guidance, suggestions and encouragement, encouragement at every step, and continuous assistance right from conceptualization of the project work for the preparation of this thesis.

We heartily thank **Mrs. T. Lakshmi, HOD of Pharm- D,** my research guide for her tolerance, keen interest, valuable guidance, logical thoughts, kind co-operation, constructive criticism, and encouragement in every step, and continuous assistance right from conceptualization of the project work to the preparation of this thesis and suggestions which helped me to complete every aspect of this project work.

Just as a music conductor is important for an orchestra, the Principal of a college is pivotal. My principal, teacher **Dr. T. Rama Rao, Associate Director and Principal,** never went back in taking personal interest, providing constant encouragement and valuable advice. I am deeply thankful for facilitating the project and creating a conductive environment for completing the project.

We immensely thankful to **Dr. B Soma Raju, Chairman and Managing Director, Care Hospitals,** for his rejuvenating inspiration, valuable suggestions and encouragement given to me during the project work.

We would like to express my deep sense of gratitude to **Dr. N Krishna Reddy, Vice Chairman,** for his guidance, valuable suggestions and affection during my course.

*We immensely thankful to **Dr. Raghava Raju, Medical Director**, for his kind co-operation, inspiring guidance, supervision and help in completing this work.*

*We would like to express my sincere gratitude and feels immense pleasure in thanking to **Dr. Anuj Kapadiya, MD, DM, Director of Cath Lab services**, for his guidance during the term of my project. Without his valuable assistance, this work would not have been completed.*

*We profusely thank **Dr. Riyaz Khan, Facility chief operating officer,** for the infrastructure and all other essential facilities and encouragement given to us during the project work.*

*We take the privilege to express my heart full gratitude to **Dr. Gopi Krishna, Medical Superintendent, Dr. S. Naga Satish Kumar, Associate Medical Superintendent and Dr. Harish Jawalkar, Associate Medical Superintendent** for their co-operation, affection, encouragement and moral support throughout our project.*

*We wish to extend my sincere thanks to **Dr. Naresh, DNB Cardiologist** for encouragement, timely help in my project work and support.*

*We express my sincere thanks to **Mr. Venkatesh, Clinical Pharmacist and Research Coordinate, and Mr. Vidya Sagar, Clinical Pharmacist** for their cooperation and help in completing this work.*

*It is my privilege to express thanks to **Dr. J. N. Narendra Sharath Chandra** for encouragement and moral support throughout the project.*

*Our sincere thanks to **Mr. Sudheer** for his help and guidance in performing statistics of the study.*

*We would be failing in my duty if I don't acknowledge the help of **Authors** of journals and books, the sentinels of my project work.*

*Our sincere thanks to senior's and my entire batch mates, for all the love, support they have given throughout my **Pharm D** and letting me be me and realize my potential.*

*Finally, yet importantly, we thank all the **Patients** who participated in the study without whom the study would not be possible.*

Thanks again......

By,

Chirumamilla Murali krishna
Thanneru Keerthi
Marupakala Shirisha

Dedicated to my beloved Parents, Teachers and Friends

Thank you. Without your support and patience, I would have never achieved my dream

Abbreviations

Abbreviated Form	Full Form
MI	Myocardial Infarction
AMI	Acute Myocardial Infarction
HCN	Hyperpolarization-Activated Cyclic-Nucleotide-gated channels
LVEF	Left Ventricular Ejection Fraction
SAN	Sino Atrial Node
I_f	Funny Current Channel
HR	Heart Rate
HF	Heart Failure
BP	Blood Pressure
CAD	Coronary Artery Disease
CVD	Cardiovascular Disease
cTn	Cardiac Troponin
ECG	Electro Cardiography
ACS	Acute Coronary Syndrome
UA	Unstable Angina
STEMI	ST - Elevated Myocardial Infarction
NSTEMI	Non ST - Elevated Myocardial Infarction
PCI	Precutaneous Coronary Intervention
CABG	Coronary Artery Bypass Grafting
BNP	Brain Natriuretic Peptide
NT - ProBNP	N – Terminal Pro b-type Natriuretic Peptide
LDL	Low Density Lipoprotein
HDL	High Density Lipoprotein
DM	Diabetes Mellitus
HTN	Hypertension
BMI	Body Mass Index
tPA	Tissue Plasminogen Activator
APTT	Activated Partial Thromboplastin Time
ACE Inhibitors	Angiotensin Converting Enzyme
RV	Right Ventricular
LV	Left Ventricular
AWMI	Anterior Wall Myocardial Infarction
PWMI	Posterior Wall Myocardial Infarction
IWMI	Inferior Wall Myocardial Infarction
LWMI	Lateral Wall Myocardial Infarction
RBBB	Right Bundle Branch Block
LBBB	Left Bundle Branch Block
CYP3A4	Cytochrome P450 3A4

Table of Contents

List of Tables

List of Figures

Chapter-1

Introduction

Myocardial infarction (MI) is the term used for an event of Heart attack which is due to blockage of the arteries resulting in absent of blood flow to the myocardium and causing death of muscle called as infarction.[1] Myocardial necrosis due to myocardial ischemia is defined as myocardial infarction. MI requires cardiac myocyte necrosis with an increase and/or an decrease in plasma of cardiac troponin (cTn). At least one cTn measurement should be greater than the 99th percentile normal reference limit during:

- symptoms of myocardial ischemia;
- new (or presumably new) significant ECG ST-segment/T-wave changes or left bundle branch block;
- the development of pathological electrocardiographic (ECG) Q waves;
- new loss of viable myocardium or regional wall motion abnormality identified by an imaging procedure; or
- identification of intracoronary thrombus by angiography or autopsy.[2]

Reasons for the elevation of cardiac troponin values because of myocardial injury. (shown in Table - 1).[3]

Table No- 1: Myocardial Injury

Myocardial injury related to acute myocardial ischemia
Alterosclerotic plaque disruption with thrombosis.
Myocarial injury relate to acute myocardial ischemia because of oxygen supply/demand imbalance
Reduced myocardial perfusion, eg.
Coronary artery spasam, microvascular dysfunction
Coronary embolism
Coronary artery dissection
Sustained bradyarrhythmia
Hypotnsion or shock
Respiratory failure
Severe anaemia
Increased myocardial oxygen demand, eg.
Sustained tacyarrhythmia
Severe hypertension with or without left ventricular hypertrophy
Other causes of myocardial injury
Cardiac conditions, eg.

Heart failure
Myocarditis
Cardiomyopathy (any type)
Takostubo syndrome
Coronary revasculariztion procedure
Coronary procedure other than revasculariztion
Catheter ablation
Defibrillator shocks
Cardiac confusion
Systemic conditions, eg.
Sepsis, infectious disease
Cronic kidney disease
Stroke, subarachnoid haemorrhage
Pulmonary embolism, pulmonary hypertension
Infiltrative diseases, e.g. amyloidosis, sarcoidosis
Chemotherapeutic agents
Critically ill patients
Strenuous exercise

Myocardial Infarction combines clinical symptoms, cardiac biomarkess and electrocardiogram (ECG) changes. Detection of a rise and a fall of troponin, expressed in ng/L or pg/mL, is essential to the diagnosis of acute MI[2]

The classification distinguishes between type-1 myocardial infarction due to thrombosis of an atherosclerotic plaque and type-2 myocardial infarction due to myocardial oxygen supply-demand imbalance in the context of another acute illness. Myocardial infarctions presenting as sudden death (type-3), or after percutaneous coronary intervention (type-4) and coronary artery bypass grafting (type-5).[4]

Classification of MI:

Universal Definition of Myocardial Infarction, recently classified myocardial infarction to five different subtypes. (shown in Figure - 1).[4]

Type 1 infarction occurs because of plaque rupture, ulceration or dissection, etc. in the presence of unstable atherosclerotic coronary artery disease (shown in Figure – 3). that comprises blood flow with resultant myocardial necrosis. (shown in Figure – 2).[3]

 **TYPE 1 MYOCARDIAL INFARCTION**
Spontaneous myocardial infarction related to ischaemia due to a primary coronary event such as plaque erosion and/or rupture, fissuring or dissection

 **TYPE 2 MYOCARDIAL INFARCTION**
Myocardial infarction secondary to ischaemia due to either increased oxygen demand or decreased supply

 **TYPE 3 MYOCARDIAL INFARCTION**
Sudden unexpected cardiac death often with symptoms suggestive of myocardial ischaemia

 **TYPE 4 MYOCARDIAL INFARCTION**
Myocardial infarction associated with percutaneous coronary intervention (4a) or stent thrombosis (4b)

 **TYPE 5 MYOCARDIAL INFARCTION**
Myocardial infarction associated with cardiac surgery

 **MYOCARDIAL INJURY**
Multifactorial aetiology; acute or chronic based on change in cardiac troponin concentrations with serial testing

Figure No- 1: Myocardial Infarction Types

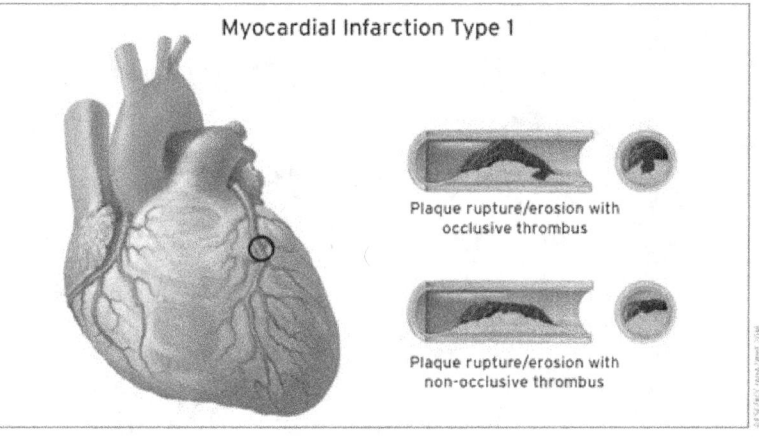

Myocardial Infarction Type 1

Plaque rupture/erosion with occlusive thrombus

Plaque rupture/erosion with non-occlusive thrombus

Figure No- 2: Myocardial Infarction Type-1

Type 2 infarction is caused by a disequi- librium between oxygen supply and demand produced by factors other than unstable coronary artery disease (CAD) i.e. toxic effects of endogenous circulating compounds—catecholamines, endothelial dysfunction, etc. (Shown in Figure – 4).[3]

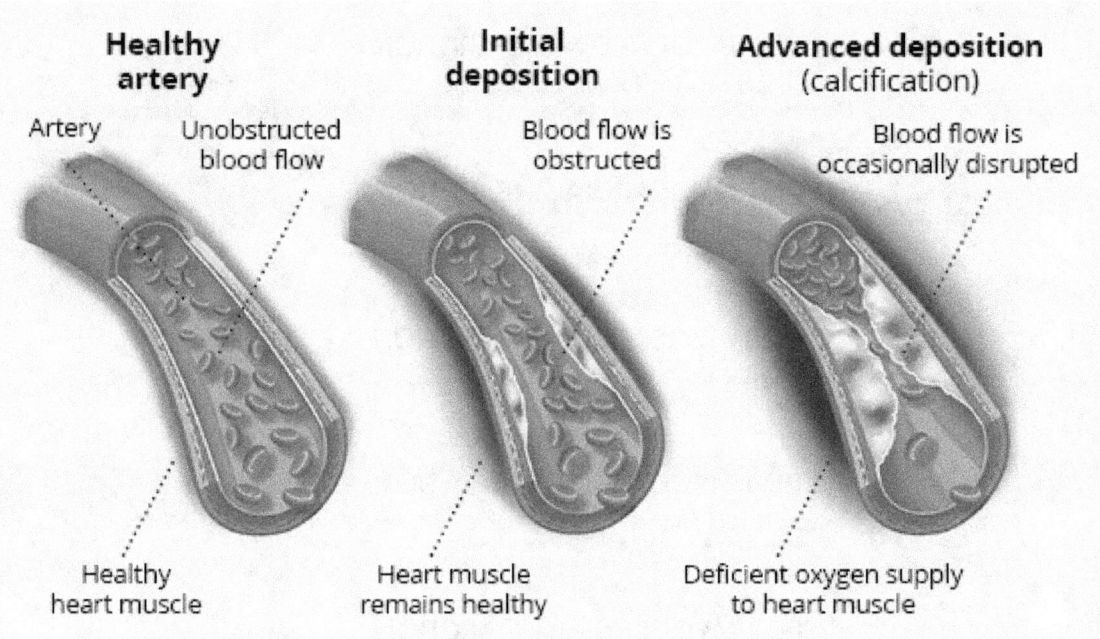

Figure No-3: Progression of Atherosclerosis

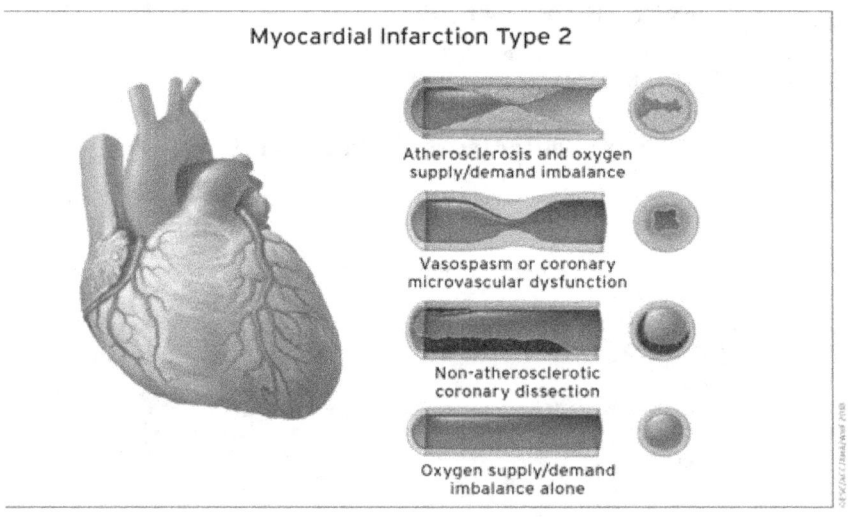

Figure No- 4: Myocardial Infarction Type-2

Type 3 infarction involves patients with cardiac death resulting from symptoms associated with myocardial ischemia but for whom cardiac biomarker results are lacking.[3]

Type 4 (a and b) infarction is linked to percutaneous coronary intervention and stent thrombosis, respectively.[3]

Type 5 infarction is related to coronary artery bypass grafting.[3]

Epidemiology of Myocardial Infarction:

According to 2014, based on the self-reported national survey of the UK, the prevalence of MI was reported as 640,000 in men and 275,000 in women; this represents about 915,000 people that have suffered an MI in the UK. In 2013, the prevalence of MI in men was about three timeshigher than for women in the UK [16]. As shown in Figure 2, the prevalence of age-specific MI extends from 0.06% of men.[5]

The incidence of myocardial infarction was more significant among black men (12.9/100,000 males) who are in the age group 75–84 years of age than whites for both men (9.1/ 100,000 males) and women (7.8/100,000 females). Similar trends exist in other age groups and their counterparts.[5]

Aetiology of Myocardial Infarction:

Acute myocardial infarction is caused by the complete occlusion of a coronary artery with thrombus. The thrombus occurs at the site of a plaque which has ruptured, exposing its inner core and thus promoting thrombus formation.[6]

Causes of Myocardial Infarction:

A myocardial infarction, means that some portion of the heart muscle has died. Myocardial infarction almost always happen when the blood supply to the heart muscle has been cut off. In most cases, a Myocardial infarction is an acute event, resulting from the sudden rupture of an atherosclerotic plaque in the wall of a coronary artery, in a person with typical coronary artery disease (CAD). However, there are other conditions that can also produce a myocardial infarction.[7]

Acute Coronary Syndrome, Coronary Artery Spasm, Microvascular Angina, Stress Cardiomyopathy, Viral Myocarditis, Blood Clotting Disorders, Coronary Artery Embolism, Genetics and Myocardial Infarction, Other causes are due to – Bad Cholesterol, Saturated fats and Trans fat.[7-8]

Signs and symptoms of Myocardial Infarction:

Myocardial infarction symptoms includes:

Chest pain, Radiating pain(left arm or shoulder, right arm or shoulder, both arms or shoulders, neck, back, epigastric), Oppressive pain, Nausea/Vomiting, Sweating, Absence of chest-wall tenderness on palpitation(absence of tenderness), Shortness of breath, Anxiety, Dizziness, Fast heart rate.[9]

Diagnosis of Acute Myocardial Infarction :

AMI is diagnosed in the emergency based on ST segment elevation of more than 1.5 mm in 2 or more leads. Physical examination, electrocardiogram and cardiac enzymes have conventionally been used for diagnosis of AMI. Echocardiography in the emergency room is being used more and more frequently with greater accuracy for the same.[10]

Electrocardiogram: ECG is generally the first investigation available for making a diagnosis in a patient presenting with acute severe chest pain.

Echocardiography: Electrocardiography is helpful in the evaluation of chest pain, especially during active chest pain.

Biomarkers: Cardiac biomarkers have conventionally being used for diagnosis of acute myocardial infarction. Common biomarkes used are – Myoglobin, Creatine Kinase, Creatine Kinase Isoenzymes, Cardiac specific Troponins.

BNP and NT-ProBNP: Although originally BNP and NT-ProBNP were considered biomarkers for heart failure only, now they are also considered biomarkers of myocardial ischemia.[10]

Risk factors of Myocardial Infarction:

There are various risk factors of AMI. Among them, some are modifiable (treatable) and others are non-modifiable (cannot be changed). The major risk factors of AMI are:[11]

Physical activity:

Inactive people with multiple cardiac risk factors are morelikely to develop AMI. Physical activity may contribute up to 20%-30% reduced risk of coronary heart disease. Different types of phys-ical activities may have different effects on the risk of car-diovascular disease (CVD) and may interact together.[11]

Smoking:

Smoking is considered to be strong risk factor of myocardial infarction, premature atherosclerosis and sudden cardiac death. Smoking results in early STEMI especially in other-wise healthier patients. Cigarette smoking increases the risk for AMI by multiple and complex mechanisms.[11]

Alcohol Consumption:

Alcohol consumption is associated with an acutely higher risk of myocardial infarction in the subsequent hour among people who do not typically drink alcohol daily. There is consistent evidence that moderate habitual alcohol consumption is associated with a lower risk of cardiovascular events in subsequent months and years and that heavy episodic (binge) drinking is associated with higher cardiovascular risk.[11]

Dyslipidemia:

Increased triglyceride levels and dense, small LDL particles act as predisposing risk factors for myocardial infarction.[11]

Diabetes Mellitus:

Diabetes Mellitus have several risk factors in common with coronary artery disease(CAD), such as age, hypertension, dyslipidemia, obesity, physical inactivity and stress, an increase in the prevalence of diabetes indirectly implicates an escalating risk of CAD as well.[11]

Hypertension:

Both systolic and diastolic hypertension increase the risk of a myocardial infarction and the higher the pressure, the greater the risk. It is major risk factor of causing athero-sclerosis in coronary blood vessels, result in heart attack or myocardial infarction.[11]

Obesity/BMI:

Increased BMI is directly related to incidence of myocardial infarction. Infarction is greatly enhanced by extreme obesity because it is a recognized risk factor for myocardial infarction.[11]

Stress:

Chronic life stress, social isolation and anxiety increase the risk of heart attack and stroke. Acute psychological stress also is associated with increased risk for coronary heart disease.[11]

Gout:

Patients with gout have an increased risk of myocardial infarction. In gout patients, the inflammatory response associated with gout plays a key role in the initiation and progression of atherosclerosis, and promotion of a pro-thrombotic environment that leads to acute coronary events such as angina or myocardial infarction.[11]

Periodontal Diseases:

Periodontal diseases are a group of inflammatory diseases in which bacteria and their by-products are the principal aetiologic agents. It indicates an association between periodontal disease and acute myocardial infarction.[11]

Family history:

Family history of myocardial infarction is an independent risk factor for AMI. Several genetic variants are associated with increased risk of AMI and family history of AMI in a first-degree relative doubles AMI risk.[11]

Age:

Advanced age is associated with an increased mortality in acute myocardial infarction. The mechanism by which increasing age contributes so dramatically to mortality is unknown. About 80% of heart disease deaths occur in people aged 65 or older.[11]

Gender:

Men tend to have heart attacks earlier in life than women. Women's rate of heart attack increases after menopause but does not equal men's rate. Even so, heart disease is the leading cause of death for both men and women.[11]

Treatment:

Aim of treatment

Early treatment aims to reduce the extent of myocardial damage. As the myocardium is damaged by a diminished oxygen supply due to the obstructed coronary artery, infarct size can be reduced in two ways:

- Dissolution of the thrombus to restore coronary blood flow.
- Decreasing myocardial oxygen consumption.[12]

Initial treatment

The immediate treatment of myocardial infarction include aspirin, which prevents blood from clotting and Nitro-glycerin to treat angina(Chest pain). To tackle the arterial blockage of blood clot injection such as tissue plasminogen activator, streptokinase or fibrinolytic are used.[13]

Aspirin

All patients with a suspected myocardial infarction should be given aspirin. It is a powerful antiplatelet drug, with a rapid effect, which reduces mortality by 20%. Aspirin, 150-300 mg.

Fibrinolytic therapy

The mainstay of treatment is fibrinolytic therapy. This is given to dissolve the thrombus in the artery and restore flow. The two fibrinolytic drugs commonly used are streptokinase and tissue plasminogen activator (tPA).[12]

Streptokinase

Streptokinase produces generalised systemic fibrinolysis. An intravenous infusion of 1.5 million units is given over 30-60 minutes. Most patients will develop hypotension if streptokinase is given quickly, but this is usually easily overcome by slowing the infusion and giving fluid.

Tissue plasminogen activator (tPA)

As tPA specifically binds to thrombus, it produces local fibrinolysis. It does not have the same systemic effects as streptokinase.

Bolus – 15mg Maintenance - 0.75 mg/kg over 30 minutes (not to exceed 50 mg) then 0.5 mg/kg over 60 minutes (not to exceed 35mg).[12]

Heparin

Heparin is an antithrombin agent. It has been utilised with both fibrinolytic drugs and given subcutaneously and intravenously.

Heparin and streptokinase

Intravenous heparin is given as a 5000 unit bolus followed by 1000 units per hour intravenously, adjusted after 24 hours according to the activated partial thromboplastin time (APTT). APTT measurements are little use in the first 24 hours as streptokinase also raises the APTT.[12]

Heparin and tPA

Currently, it is believed that heparin should be given with tPA. The standard regimen is an initial bolus of 5000 units, followed by an infusion of 1000 units per hour adjusted after 6 hours for APTT.

ACE inhibitors

ACE inhibitors reduce the mortality of myocardial infarction. ACE inhibitors are given only to patients with large infarcts and those with clinical signs of left ventricular failure. Captopril 6.25 mg, or equivalent low doses of another ACE inhibitor, should be used as a first dose and, if tolerated, the dose increased to at least 25 mg twice daily of captopril or the equivalent dose of the alternatives.[12]

Beta blockers

Intravenous beta blockers such as atenolol, metoprolol and timolol reduce the incidence of arrhythmias, infarct size and mortality. As the effect is relatively small, they are not widely used. Beta blockers can be given if the patient is haemodynamically stable with a heart rate above 50 beats per minute and systolic blood pressure above 100 mmHg. The standard regimen is atenolol 5 mg intravenously over 5 minutes followed 10 minutes later by a further 5mg. Oral beta blockade is commenced 30 minutes later. Many centres use only oral beta blockade (atenolol 50mg, metoprolol 50 mg) commenced as soon as possible after admission.[12]

Glyceryl trinitrate

Intravenous glyceryl trinitrate reduces preload and after load and may help to keep the coronary vessels open. In small studies, intravenous glyceryl trinitrate for 24 hours reduced mortality, but this has not been confirmed by large trials. Intravenous glyceryl trinitrate can be used routinely or only for ongoing chest pain or left ventricular failure (the standard dose is 5 microgram/minute and this can be titrated against the blood pressure).[12]

Complications :

Complications of MI include the development of congestive heart failure (systolic and diastolic), pericardial effusion, left ventricular (LV) thrombus, right ventricular (RV) infarction, LV aneurysm, LV pseudoaneurysm, free wall rupture, ventricular septal rupture, mitral regurgitation, and dynamic LV outflow tract obstruction causing hemodynamic collapse.[14]

Among the complications noted during hospital stay the commonest were acute pulmonary edema, cardiogenic shock and arrhythmias seen in 18%, 16% and 13% patients respectively. Arrhythmias noted were varying degrees of heart block (8%), atrial fibrillation (3%) and ventricular tachycardia (2%).[15]

Prevention:

Primary Prevention - primary prevention can be basically achieved trough a reduction in high blood pressure and by correcting dyslipidemia.

Secondary Prevention – Secondary prevention of MI can be again obtained by controlling blood pressure and reducing serum cholesterol in patients surviving acute MI who can also benefit from the administration of t3-blockers, aspirin and probably ACE inhibitors particularly in presence of left ventricular dysfunction.[16]

Life Style Modifications:

Diet and lifestyle recommendations to prevent MI

Consume an overall healthy diet, Aim for a healthy body weight, aim for recommended levels of LDL and HDL cholesterol and triglycerides, aim for a normal BP, aim for a normal blood glucose level, be physically active, avoid use of and exposure to tobacco products.

Dietary changes

Less than 7% of the day's total calories from saturated fat, 25% to 35% of the day's total calories from fat, Less than 200 mg of dietary cholesterol a day, Limit sodium intake to 2,400 mg a day and Just enough calories to achieve or maintain a healthy weight and reduce blood cholesterol levels.[17]

Acute Myocardial infarction has been divided into ST elevation and non–ST elevation Myocardial Infarction. Acute coronary syndrome are usually caused by the rupture of plaque in a coronary artery, leading to either partial or complete obstruction of the vessel. [18-19]

Acute coronary syndromes lead to myocyte injury and subsequent death. In the clinical setting, their classification is based upon electrocardiographic presentation; ST-segment elevation myocardial infarction (STEMI; ≥ 2 mm ST segment eleva- tion and prominent T waves on the electrocardiogram) and non-ST-segment elevation myocardial infarction (NSTEMI; symptoms of acute coronary syn- drome exemplified by ST-segment depression and T wave inversion on the elec- trocardiogram). Location of ST-segment changes often depends on the myocar- dial region affected by acute ischemia.[20]

ACS comprises non ST elevation ACS (NSTE-ACS), Unstable angina (UA) and ST elevation MI(STEMI).(shown in Figure – 5).[21]

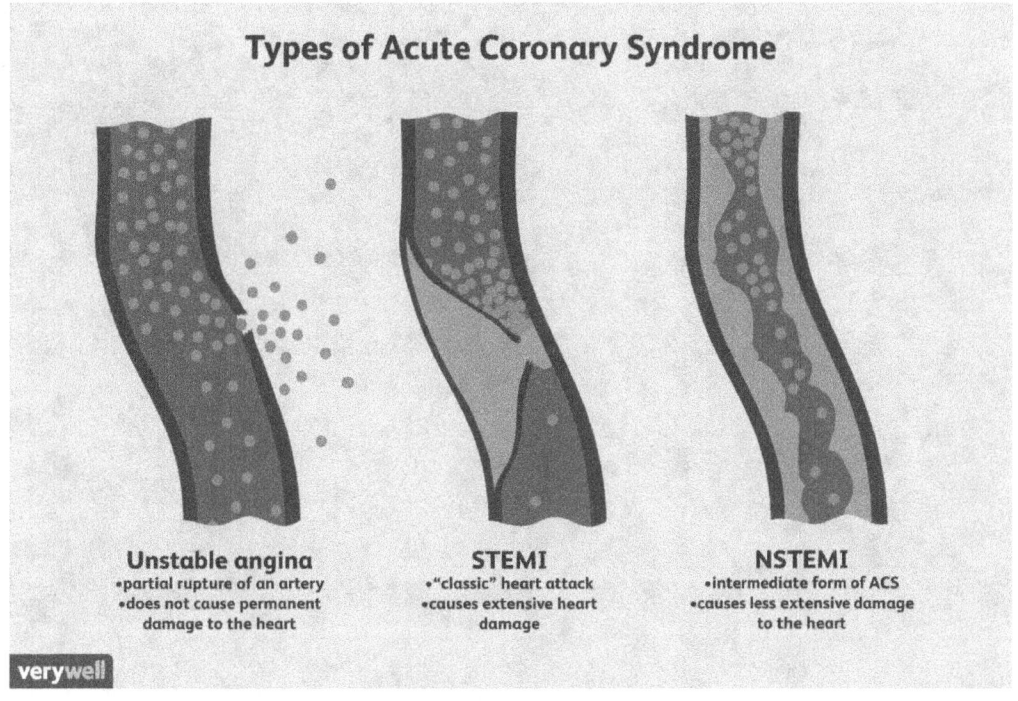

Figure No- 5 Types of Acute Coronary Syndrome

STE-ACS (ST Elevation Acute Coronary Syndrome) is defined as an acute coronary syndrome with ST elevations on ECG. STE-ACS has been discussed previously (refer to STEMI – ST Elevation Myocardial Infarction and ST Elevations in Ischemia and Infarction). Virtually all patients with STE-ACS develop myocardial infarction, which is then classified as STEMI (ST Elevation Myocardial Infarction). (shown in Figure – 6).[22]

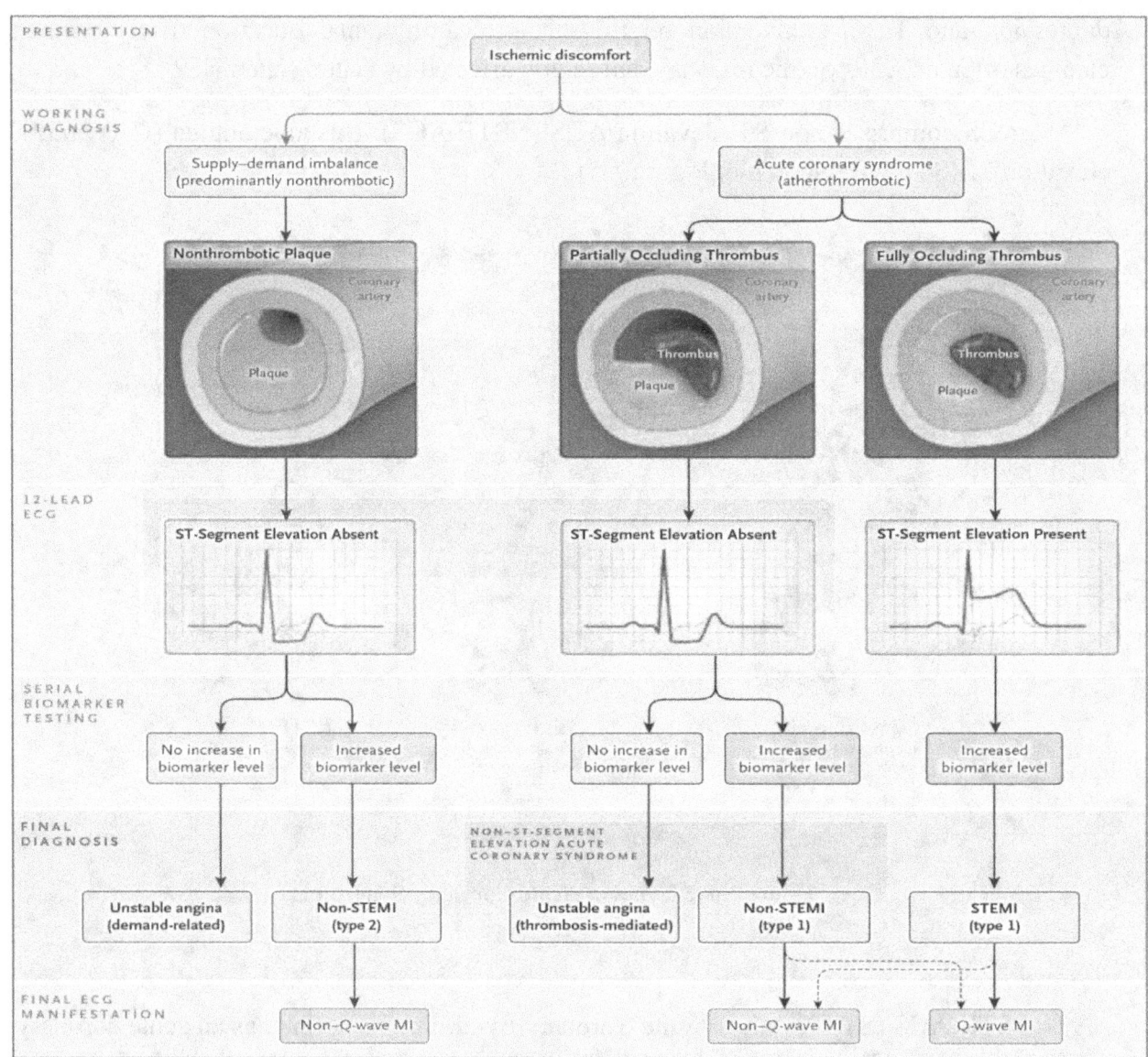

Figure No- 6: Spectrum of Pathologic and Clinical ST-Segment Elevation Acute Myocardial Infraction (STEMI) and Non-STEMI Acute Coronary Syndrome

ECG- electrocardiogram, and MI- myocardial infraction

The majority of patients with NSTE-ACS will exhibit elevated troponins, which is evidence for myocardial infarction and therefore defines the condition as NSTEMI (Non ST Elevation Myocardial Infarction). Cases who do not display elevated troponins are classified as unstable angina (UA). The vast majority of patients with NSTEMI or unstable angina present with ST segment depressions and/or T- wave inversions on ECG. (shown in Figure – 7).[22]

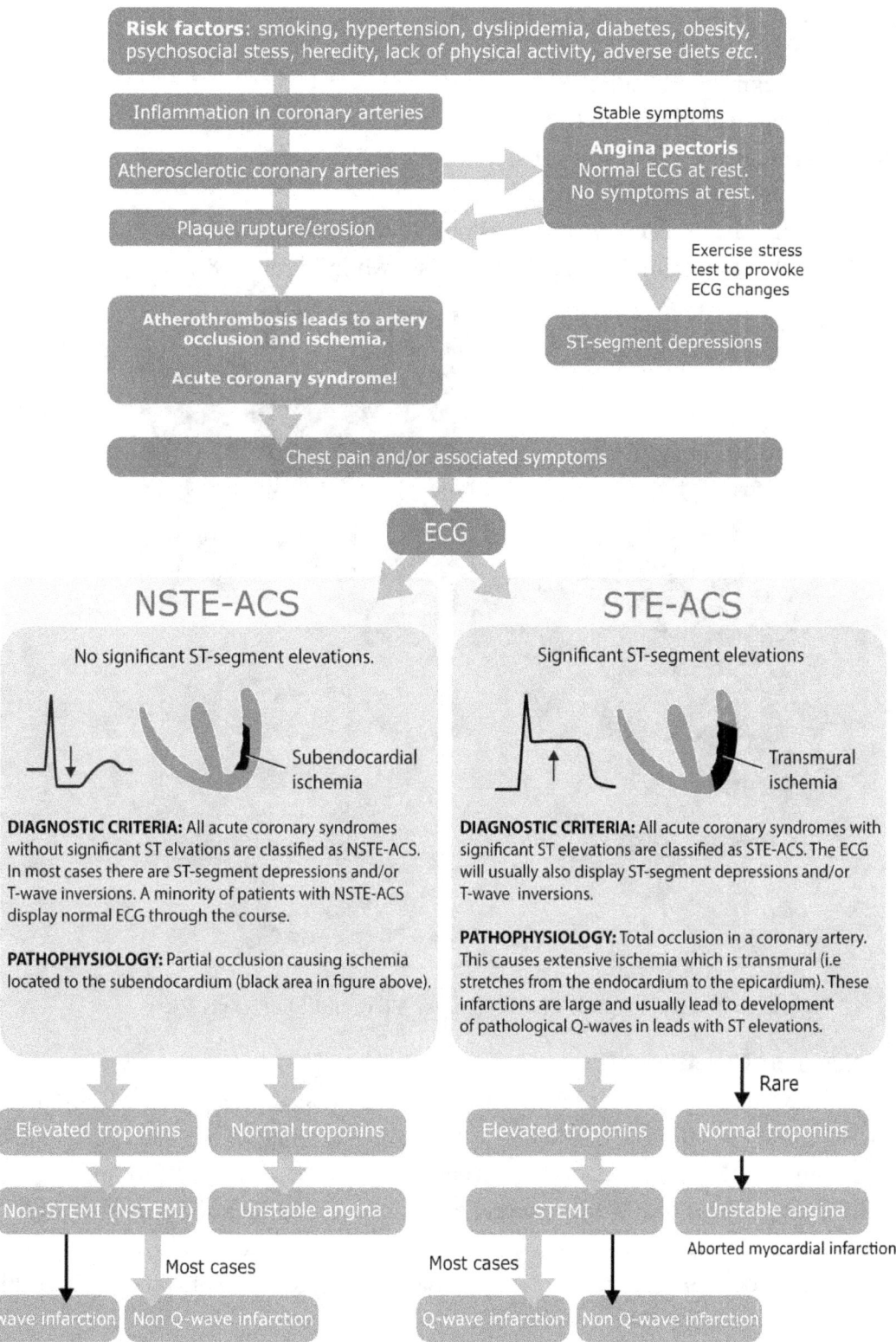

Figure No- 7: Myocardial Infarction (Q- wave infarction an Non Q-wave infarction)

Types of STEMI : [23]

1. Anterior ST Segment Elevation MI
2. Inferior ST Segment Elevation MI
3. Posterior ST Segment Elevation MI
4. Anterolateral ST Segment Elevation MI
5. Acute MI with a Right Bundle Branch Block
6. Acute MI with a Left Bundle Branch Block (shown in Figure – 8).

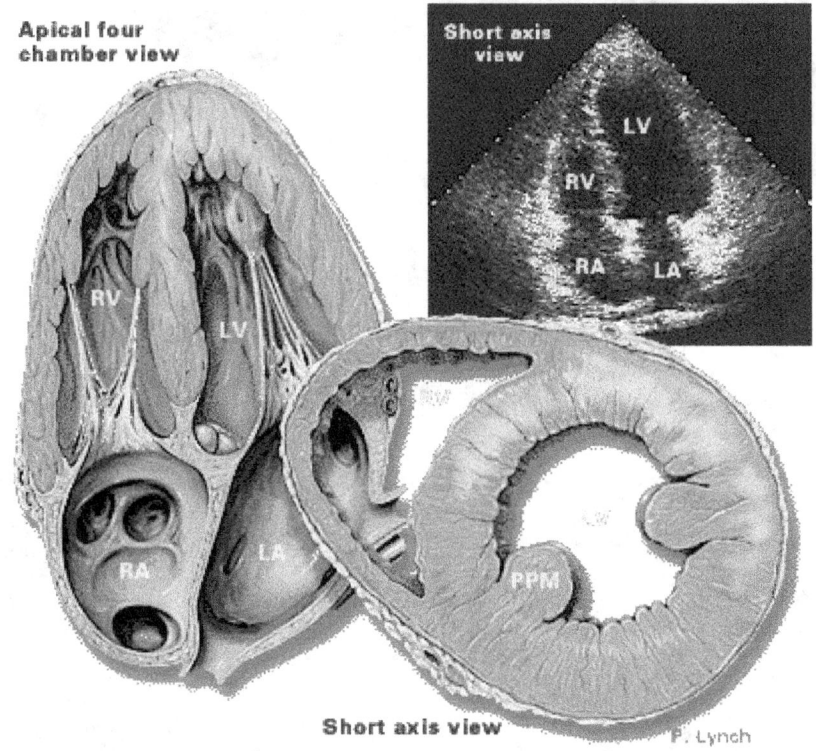

Figure No- 8: Heart Apical Four Chamber View and Short Axis View

Anterior ST Segment Elevation MI :

An anterior wall myocardial infarction — also known as anterior wall MI, or AWMI, or anterior ST segment elevation MI, or anterior STEMI — occurs when anterior myocardial tissue usually supplied by the left anterior descending coronary artery suffers injury due to lack of blood supply. When an AWMI extends to the septal and lateral regions as well, the culprit lesion is usually more proximal in the LAD or even in the left main coronary artery. This large anterior myocardial infarction is termed an extensive anterior. (shown in Table – 1). [24]

The ECG findings of an acute anterior myocardial infarction wall include:

1. ST segment elevation in the anterior leads (V3 and V4) at the J point and sometimes in the septal or lateral leads, depending on the extent of the MI. This ST segment

elevation is concave downward and frequently overwhelms the T wave. This is called tombstoning" for obvious reasons; the shape is similar to that of a tombstone.

2. Reciprocal ST segment depression in the inferior leads (II, III and aVF).(shown in Figure – 9). [24]

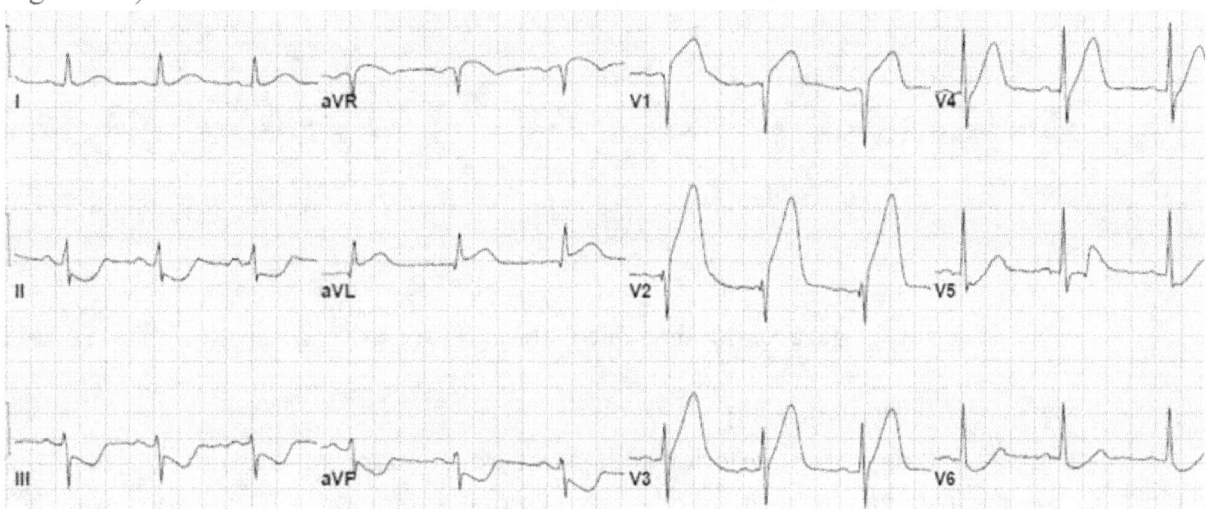

Figure No-9: Anterior ST Elevated MI ECG

Inferior ST Segment Elevation MI:

An inferior wall myocardial infarction — also known as IWMI, or inferior MI, or inferior ST segment elevation MI, or inferior STEMI — occurs when inferior myocardial tissue supplied by the right coronary artery, or RCA, is injured due to thrombosis of that vessel. When an inferior MI extends to posterior regions as well, an associated posterior wall MI may occur. (shown in Table – 2). [25]

The ECG findings of an acute inferior myocardial infarction include the following:

1. ST segment elevation in the inferior leads (II, III and aVF)

2. Reciprocal ST segment depression in the lateral and/or high lateral leads (I, aVL, V5 and V6). Shown in Figure – 10). [25]

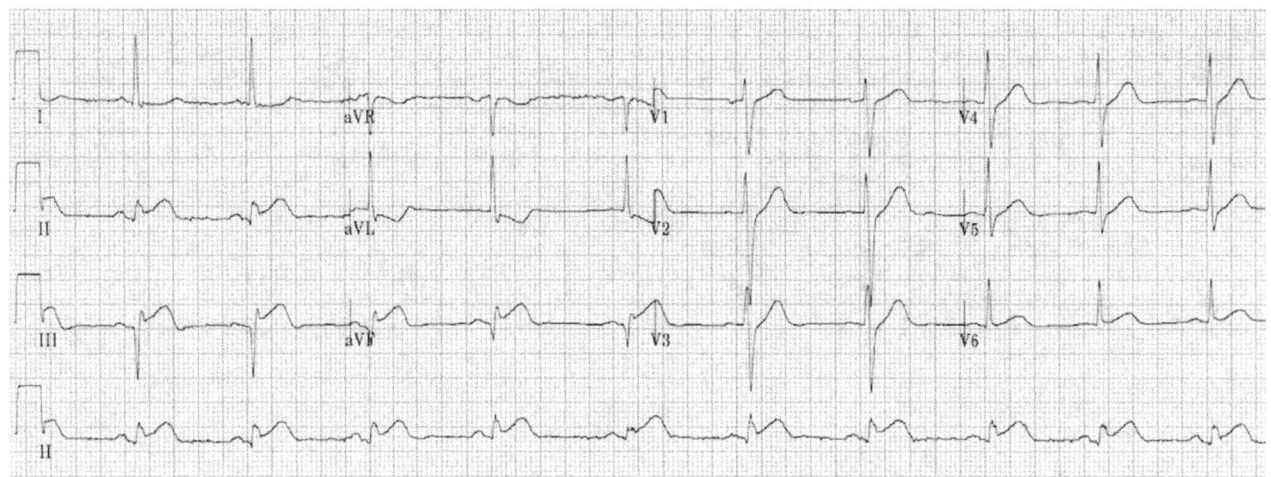

Figure No- 10: Inferior ST Elevated MI ECG

Posterior ST Segment Elevation MI :

The ECG findings of a posterior wall myocardial infarction are diferent than the typical ST segment elevation seen in other myocardial infarctions. A posterior wall MI occurs when posterior myocardial tissue (now termed inferobasilar), usually supplied by the posterior descending artery — a branch of the right coronary artery in 80% of individuals — acutely loses blood supply due to intracoronary thrombosis in that vessel. This frequently coincides with an inferior wall MI due to the shared blood supply.(shown in Table – 2).[26]

The ECG findings of an acute posterior wall MI include the following:

1. ST segment depression (not elevation) in the septal and anterior precordial leads (V1-V4). This occurs because these ECG leads will see the MI backwards; the leads are placed anteriorly, but the myocardial injury is posterior.
2. A R/S wave ratio greater than 1 in leads V1 or V2.
3. ST elevation in the posterior leads of a posterior ECG (leads V7-V9). Suspicion for a posterior MI must remain high, especially if inferior ST segment elevation is also present.
4. ST segment elevation in the inferior leads (II, III and aVF) if an inferior MI is also present. (shown in Figure – 11). [26]

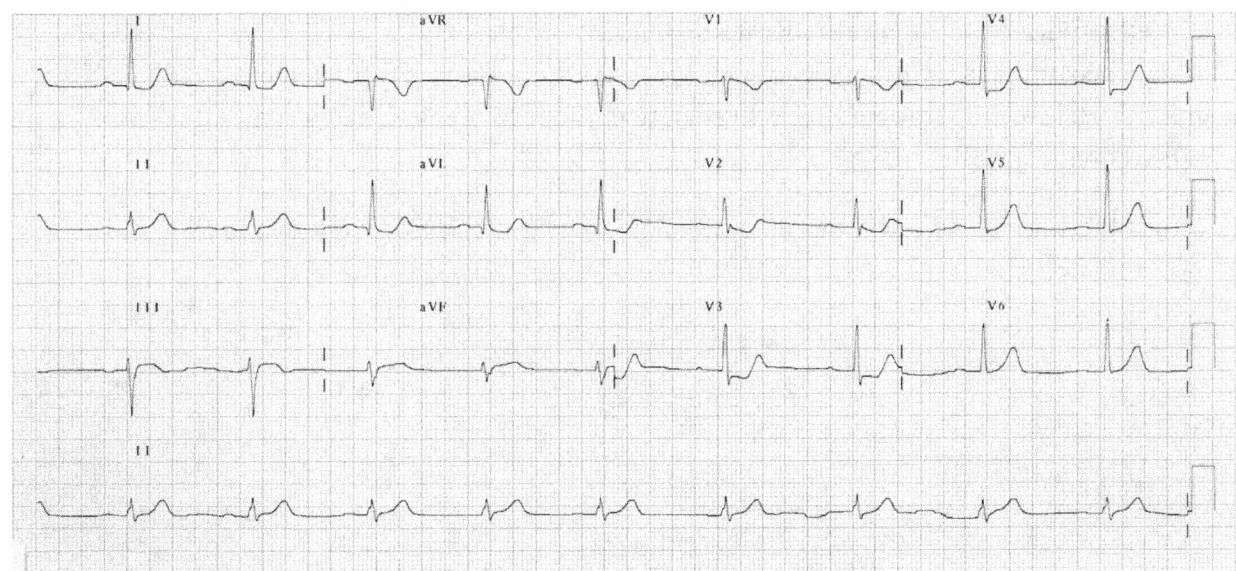

Figure: 12 Posterior ST Elevated MI ECG

Figure No- 11: Posterior ST Elevated MI ECG

Anterolateral ST Segment Elevation MI:

STE in the leads I, L and V5-6 localize the infarct to the lateral wall. This location is variable in its presentation between inferior and anterior MI, and has features that may be seen in either or both.[27] Acute anterolateral MI is recongnized by ST segment elevation in leads I, aVL and the precordial leads overlying the anterior and lateral surfaces of the heart (V3 - V6) (shown in Figure – 12).

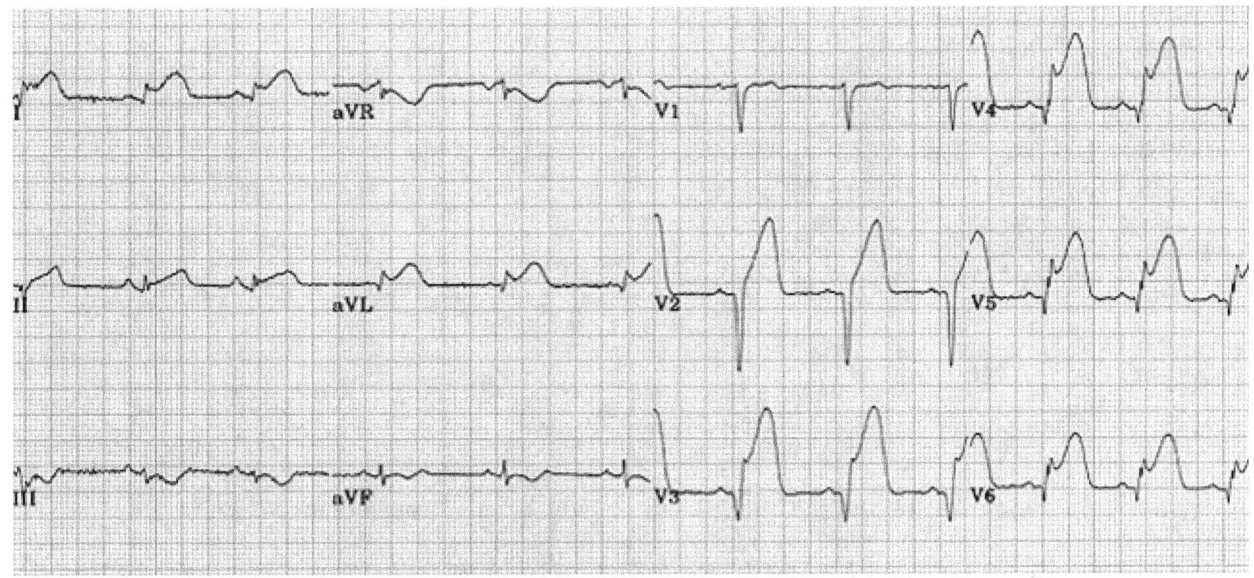

Figure No- 12: Anterolateral ST Elevated MI ECG

Acute MI with a Right Bundle Branch Block:

Recall that a right bundle branch block does not stop us from detecting a STEMI on an ECG. Below are some examples to review in order to recognize anterior and inferior STEMIs with a RBBB. [23]

Here is the anterior STEMI with a right bundle branch block ECG. (shown in Figure – 13).

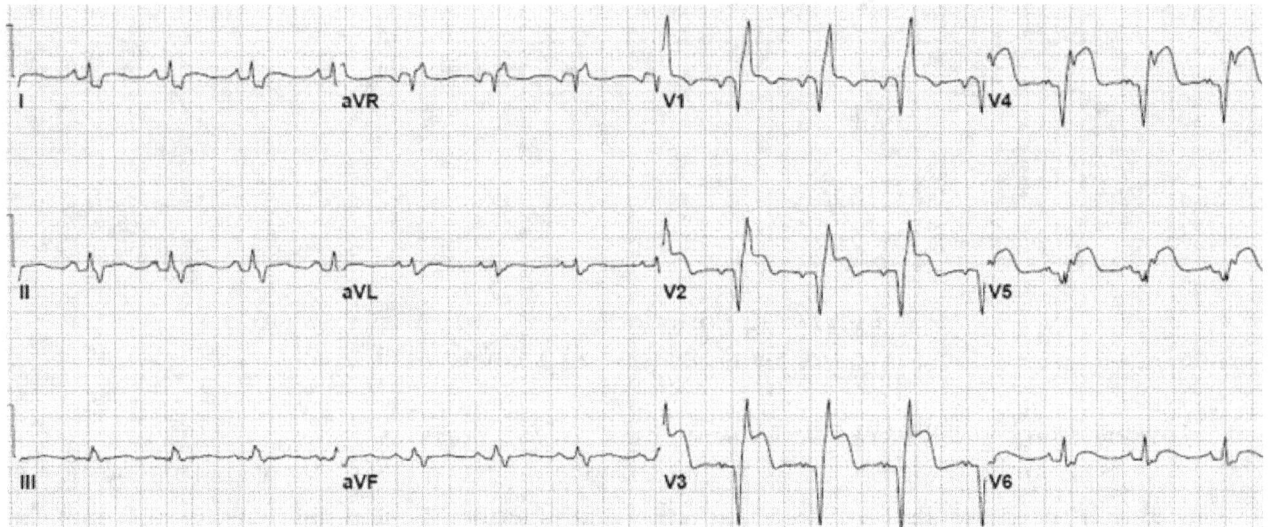

Figure No- 13: Anterior STEMI with a Right Bundle Branch Block ECG

Now, here is an inferior STEMI with a **Right Bundle Branch Block (RBBB)** on the ECG. Note the reciprocal depression in lead I and aVL.

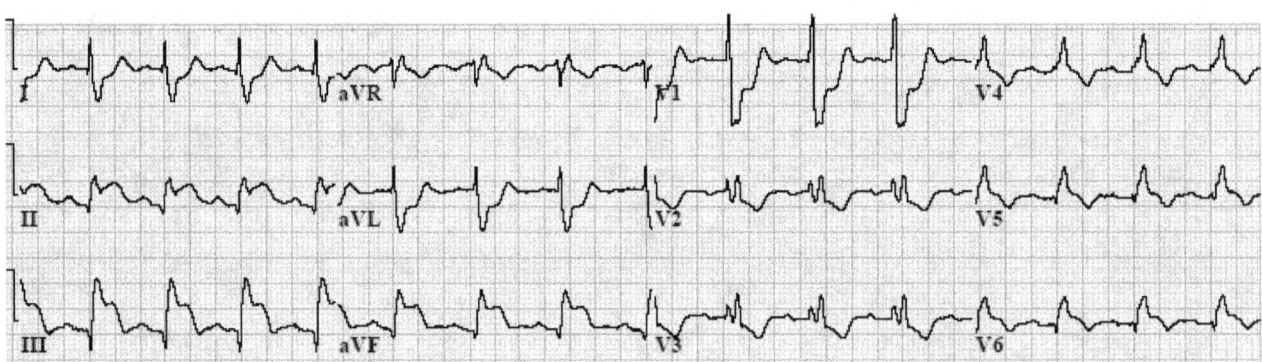

Figure No- 14: Inferior STEMI with a Right Bundle Branch Block MI ECG

Acute MI with a Left Bundle Branch Block :

Patients with a suspected myocardial infarction (MI) in the setting of a left bundle branch block (LBBB) present a unique diagnostic and therapeutic challenge to the clinician. A diagnosis of MI with electrocardiogram (ECG) is especially difficult in the setting of

LBBB because of the characteristic ECG changes caused by altered ventricular depolarization. (shown in Figure–15).[28]

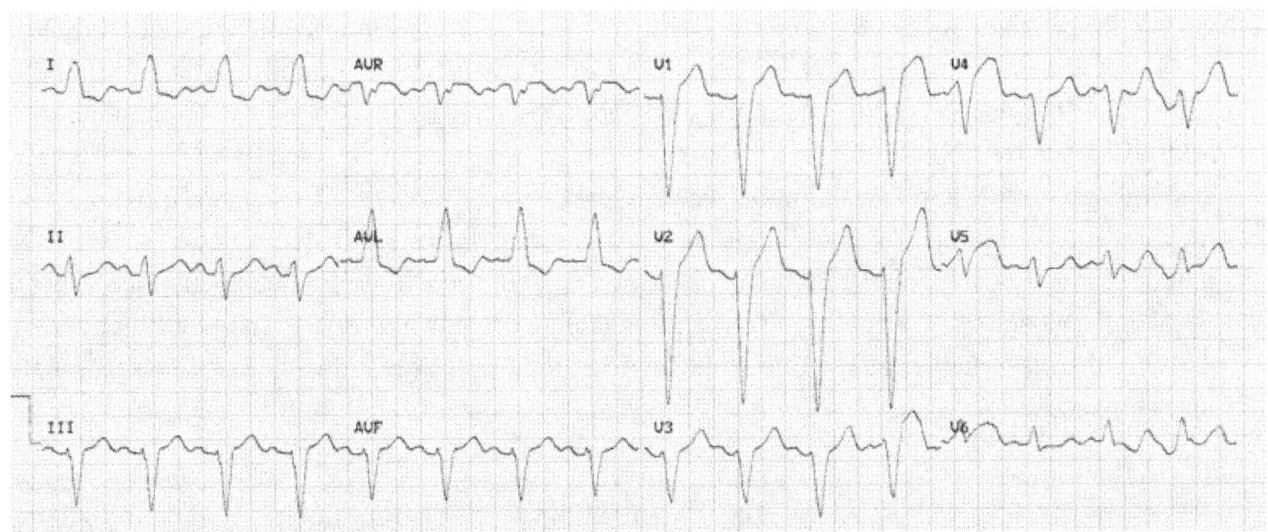

Figure No- 15: Left Bundle Branch Block MI ECG

Table No-2: Myocardial Infarction

Myocardial Infarction Location	Vessel Involved	Leads Affected	ECG Changes
Anterior heart wall	Left anterior descending artery (diagonal branch)	V2 to V4	- Poor R-wave progression - ST-elevation - T-wave inversion
Septal wall	Left anterior descending artery (septal branch)	V1 to V2	- R wave disappears - ST elevation - T-wave inversion
Inferior wall	Right coronary artery	Leads II, II and AVf	- ST elevation
Lateral wall	Left coronary artery (circumflex branch)	I, aVl, V5 and V6	- ST elevation - T wave inversion
Posterior wall	Left coronary artery (circumflex branch) Right coronary artery	V1 to V4	- Tall R waves - ST segment depression - Upright T waves

Ivabradine:

Brand Name: Corlanor

Ivabradine is a benzazepine derivative and selective HYPERPOLARIZATION-ACTIVATED CYCLIC NUCLEOTIDE-GATED CHANNELS inhibitor that lowers the heart rate. It is used in the treatment of CHRONIC STABLE ANGINA in patients unable to take BETA-ADRENERGIC BLOCKERS, and in the treatment of HEART FAILURE. Ivabradine is a Hyperpolarization-activated Cyclic Nucleotide-gated Channel Blocker. The mechanism of action of ivabradine is as a Hyperpolarization-activated Cyclic Nucleotide-gated Channel Antagonist. Ivabradine is a novel, specific HR-lowering agent that acts in SAN cells by selectively inhibiting the pacemaker If current in a dose-dependent manner by slowing the diastolic depolarization slope, and reducing HR at rest during exercise in animals and humans.[29-30]

IUPAC Name:

3-[3-[[(7S)-3,4-dimethoxy-7-bicyclo[4.2.0]octa-1,3,5-trienyl]methyl-methylamino]propyl]-7,8-dimethoxy-2,5-dihydro-1H-3-benzazepin-4-one. [30]

Molecular Formula :

$C_{27}H_{36}N_2O_5$

2D Structure :

Mechanism of Action:

The sinoatrial node is unique in that its cells have an innate ability to generate a cyclical change in their resting membrane potential, which drives it toward the threshold needed for spontaneous depolarization. This depolarization, in turn, generates repetitive, spontaneous action potentials accounting for its automaticity. This depolarization is initiated by the opening of specific ion channels that conduct a slow, inward-depolarizing mixed sodium-potassium current, referred to as the pacemaker or "funny" current (If). (shown in Figure – 16).[31]

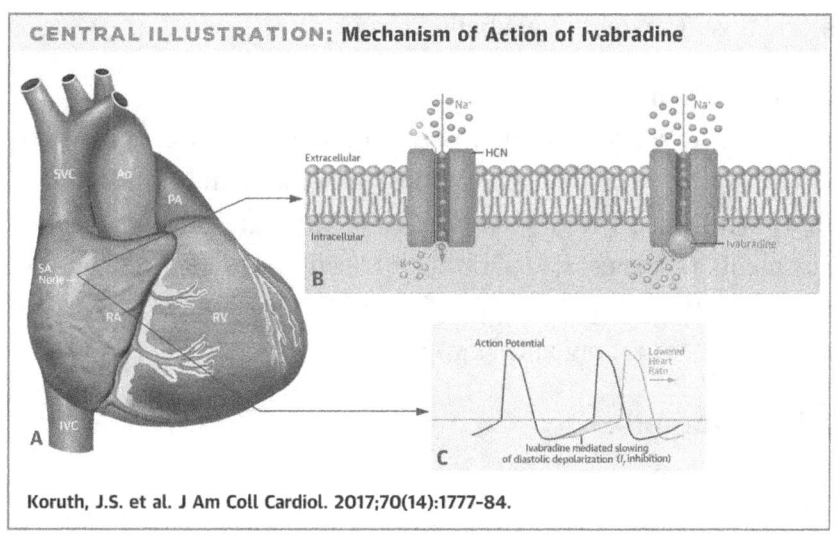

Koruth, J.S. et al. J Am Coll Cardiol. 2017;70(14):1777-84.

Figure No- 16: Ivabradine: Mechanism of Action

Pharmacodynamics:

Ivabradine is a direct, selective inhibitor of the hyperpolarization-activated cyclic-nucleotide gated funny (If) current, a mixed sodium–potassium inward channel in the sinoatrial node. The rate of sinoatrial generated impulses is dependent on spontaneous diastolic depolarization of myocytes in the sinoatrial node and the If current has a major inhibitory influence on this depolarization.[32]

Pharmacokinetics:

Absorption- Ivabradine is rapidly and almost completely absorbed after oral administration with a peak plasma level reached in approximately 1 hunger fasting condition. Food delayed absorption by approximately 1 h, and increased plasma exposure by 20 -- 30%.

Distribution- The absolute bioavailability of ivabradine is around 40%. Ivabradine is approximately 70% bound to plasma protein and the volume of distribution at steady state is close to 100 . The average plasma concentration is 10 ng/ml at steady state.

Metabolism- Ivabradine is extensively metabolized in gut and liver by oxidation through CYP3A4. The major active metabolite is the N-desmethylated derivative, and its exposure (measured by AUC) is about 40% of that of the parent compound with similar PK and PD properties. Ivabradine has low affinity for CYP3A4 and does not modify CYP3A4 substrate metabolism or plasma concentrations.

Excretion- The main elimination half-life of ivabradine is 2 h (70 -- 75% of the AUC) in plasma and an effective half-life is 11 h. The total clearance is about 400 ml/min and the renal clearance is about 70 ml/min. Metabolites are equally excreted in the feces and urine. About 4% of an oral dose is excreted unchanged in urine.[33]

Dosage and Administration :

The recommended starting dosage of ivabradine is 5 mg twice daily, administered with food. After two weeks, the patient should be assessed, and dose adjustments should be made to achieve a resting heart rate of between 50 bpm and 60 bpm (Table 2). Thereafter, further dose adjustments (if necessary) should be based on the patient's resting heart rate and tolerability. The maximum dosage of ivabradine is 7.5 mg twice daily. The recommended doses of Ivabradine are 2.5mg, 5mg and 7.5mg.[34]

Drug – Drug Interactions:

CYP-Based Interactions

Ivabradine is primarily metabolized by Cytochrome P450 3A4 (CYP3A4). The concomitant use of CYP3A4 inhibitors increases ivabradine plasma concentrations, which may exacerbate bradycardia and conduction disturbances.

Negative Chronotropes

The concomitant use of drugs that slow the heart rate (e.g., digoxin, amiodarone, and beta blockers) may enhance the bradycardic effect of ivabradine. It is important to monitor the heart rate in patients receiving ivabradine along with other negative chronotropes.

Interaction With Pacemakers

Ivabradine acts by reducing the heart rate to a target of 50 to 60 bpm. Therefore, the use of ivabradine is not recommended in patients with demand pacemakers set to a rate of 60 bpm or greater.[34]

Adverse Drug Reaction:

The most common adverse events associated with ivabradine were bradycardia, atrial fibrillation, phosphenes (luminous phenomena), and hypertension. Postmarketing information revealed that ivabradine was associated with rash, diplopia, angioedema, pruritus, urticaria, visual impairment, erythema, and vertigo.[35]

Contraindications:

Ivabradine is contraindicated in patients with acute decompensated HF, severe hepatic impairment, blood pressure below 90/50 mm Hg, a resting heart rate below 60 bpm prior to treatment, or pacemaker dependence (i.e., the patient's heart rate is maintained solely by the pacemaker). In addition, ivabradine is contraindicated in patients with sick sinus syndrome, sinoatrial block, or third-degree AV block unless a functioning demand pacemaker is present. As noted previously, the concomitant use of ivabradine and potent CYP3A4 inhibitors is also contraindicated.[34]

Dietary Instructions:

Do not eat Grape fruit or drink Grape fruit juice while taking Ivabradine.[36]

Symptoms of Overdose:

Slow heart beat, Dizziness, Excessive tiredness, Lack of energy[36]

Warning and Precautions:

Before taking Ivabradine tell your doctor if your taking antibiotics such as clarithromycin and telithromycin, antifungals such as itraconazole, certain HIV protease inhibitors such as nelfinavir and nefazodone. before taking Ivabradine tell your doctor if you have an irregular or slow heart beat, low blood pressure, a pacemaker, symptoms of heart failure that recently worsened, or liver disease.[36]

Drug Information:

Ivabradine is a specific antianginal drug. It specifically acts at I(f) channel current in the pacemaker cells of the sinoatrial(SA) node inhibitor.[37]

Ivabradine is a heart rate lowering drug, reduces the myocardial oxygen demand on exercise.Beneficial effects of Ivabradine have been demonstrated in chronic stable angina pectoris and Congestive heart failure, with optimal tolerability profile due to selective intraction with I(f) channel of sinoatrial node cells. More recently, the indication of Ivabradine has been extended for use in association with β-blockers in patients with Coronary artery disease.[38]

The pharmacokinetics and pharmcodynamic features of Ivabradine have been studied in sinoatrial node cells have shown that I(f) channel binding or unbinding Ivabradine are restricted to the open channel blockers. A peculiar feature of Ivabradine is that its blocking action is not intrinsically voltage dependent, but rather depends on the direction of ion flow across the channel pore.[39-40]

Unlike many rate lowering agents Ivabradine reduces heart rate in a dose dependent manner both at rest and during exercise. The uncommon bradycardiac effect of Ivabradine is proportional to the resting heart rate, such that the effects tends to plateau. Thus, extreme sinus bradycardia is uncommon even in octogenarians patients with increased incidence of bradycardia due to age related alteration of the sinus node.[8](Protocol Ref) Ivabradine has no direct torsadogenic potential, although, for obvious reasons the specific bradycardic drug should not be administered with agents which have known QT prolonging effects.[41]

Heart rate reduction with Ivabradine a selective and specific I(f) inhibitor, reduces myocardial oxygen demand, increases diastolic perfusion time and improves energitics in Ischemic myocardium. Ivabradine protects the myocardium during Ischemic and reduces remodeling following myocardial infarction.[42]

Role of Ivabradine in Different Cardiovascular Diseases:

Ivabradine in Acute HF:
Ivabradine should be considered for decreasing the risk of hospitalization due to HF or cardiovascular death in symptomatic patients with LVEF of 35% or less, a sinus rhythm, and resting HR of 70 bpm and above despite treatment with beta-blockers.[43]

Ivabradine in HF:
The Systolic Heart Failure Treatment With the I(f) Inhibitor Ivabradine, included only patients with HF (classes II to IV), LVEF less than 35%, sinus rhythm, and a HR greater than 70 bpm.

The starting dose of ivab- radine was 5 mg b.i.d. After a 14-day titration period, the dose was increased to 7.5 mg b.i.d. unless the resting HR was 60 bpm. If HR was between 50 and 60 bpm, the dose was maintained at 5 mg b.i.d. If the resting HR was lower than 50 bpm or if the patient had signs or symptoms related to bradycardia, the dose was reduced to 2.5 mg b.i.d.

Ivabradine in Ischemic Heart Disease:
Ivabradine is an antianginal agent because it decreases HR without a negative inotropic effect without a coronary vasocon- strictor effect. Ivabradine increases diastolic duration and coronary blood flow and preserves coronary dilation during exercise. In addition, it increases coronary flow reserve and improves collateral perfusion.

Role of Ivabradine in CHD:
Reduction in the HR in patients with CHD may help to decrease risk of myocardial ischemia. Considering the available data, ivab- radine plays a significant role in patients with CHD. Ivabradine does not have any effect on the respiratory parameters. It can be useful agents for elderly patients and those with diabetes or asthma for whom other antianginal agents (beta-blockers) are relatively contraindicated.

Ivabradine Dosage Schedule for Treatment of Angina and HF:

It is recommended that the decision to initiate or titrating the dose should be done using HR measurements. The starting dose of ivabradine should not exceed 5 mg b.i.d with meals in patients aged below 75 years. After 3e4 weeks of treatment, if the patient continues to be symptomatic, if the initial dose is well tolerated, and if resting HR remains above 60 bpm, the dose may be increased to 7.5 mg twice daily.The dose can be decreased to 2.5 mg BD if resting HR is below 50 bpm or if associated with symptoms related to bradycardia.

Use of Ivabradine Pretreatment Before Coronary Computed Tomography Angiography:

Premedication with ivabradine reduces the HR and improves the image quality of CTCA.41 Ivabradine has shown to be safe and effective in controlling HR before performing CTCA and thereby reducing the need for additional intravenous beta-blockade.

Other Potential Uses of Ivabradine:

Chronic obstructive pulmonary disease and asthma - Ivabradine can be a useful option for the prevention of increase in the HR after salbutamol inhalation in patients with COPD with coexisting CHD.50 It can also be useful in the treatment of angina pectoris and CHF in patients with CHD with COPD.Ivabradine can be added to bisoprolol in patients with ischemic heart disease with COPD, if required.

Pulmonary Hypertension:

The other potential use of ivabradine is for pulmonary arterial hypertension. The Thus, improved interactions between two ventricles and ventricular cycle events with ivabradine can have beneficial effects on pulmonary hypertension.

Sepsis and Multiple Organ Dysfunction Syndrome:

Sepsis often results in multiple organ dysfunction syndrome. Persistent tachycardia in these patients can have a deleterious ef- fect. Reduction of HR may be useful in improving survival in such cases.The effects of ivabradine in patients with multiple organ dysfunction syndrome have been studied.[43]

Study on Safety and Adverse Effects of Ivabradine in Acute Myocardial Infarction

Chapter-2

Literature Review

1. Difrancesco D, *et al.,* have done research work entitled - Heart rate lowering by specific and selective I(f) current inhibition with ivabradine; Ivabradine was selective for the I(f) current and exerts significant inhibition of that current and heart rate reduction at concentrations that do not affect other cardiac ionic currents. Ivabradine reduced heart rate without any negative inotropic or lusitropic effect, preserving ventricular contractility. In patients with left ventricular dysfunction, Ivabradine reduced resting heart rate without altering myocardial contractility.

2. Maya Guglin, *et al.,* has done research work entitled - Heart rate reduction in heart failure – Ivabradine or beta-Blockers. According to the study, Ivabradine a selective I(f) current inhibitor, decreased the heart rate in those with sinus rhythm, had been added to the most recent European guidelines on Heart failure. It was indicated in addition to β blockers in patients with decreased left ventricular ejection fraction and sinus rate of over 70bpm.

3. Foster JL, *et al.,* has done research work entitled - Ivabradine, a novel medication for treatment of heart failure with reduced ejection fraction. According to the study, Ivabradine was a safe and effective drug for Heart rate reduction to reduce hospitalizations in patients with stable, symptomatic EF (ejection fraction <35%), in sinus rhythm, and HR >70bpm

4. PG Steg, *et al.,* have done research work entitled - Safety of intravenous Ivabradine in acute ST-segment elevation MI patients treated with primary percutaneous coronary intervention: A randomised, placebo controlled, double blind, pilot study. They studied the effect of intravenous Ivabradine on heart rate. According to the study, the use of IV Ivabradine after PCI for STEMI produced a rapid and sustained reduction in heart rate, which was safe and well tolerated. IV Ivabradine may be of potential value in STEMI, by allowing rapid heart rate control without affecting BP or haemodynamics. However, to characterize its effect, further controlled trials are required to assess its impact on infaract size, LV function and clinical outcomes.

5. Irmina Urbanek, *et al.,* has done research work entitled - Risk benefit assessment of Ivabradine in the treatment of Chronic Heart failure. According to the study, if a patient cannot tolerate a β-blocker or if titration is ineffective, Ivabradine was be considered as a treatment option.

6. Jules C Hancox, *et al.,* has done research work entitled - hERG potassium channel inhibition by Ivabradine may contribute to QT prolongation and risk of torsade de points. According to the study, Ivabradine was found to be considered safe for a good Cardiac safety profile. Recent evidence has highlighted that some qualification is necessary in this regard. A meta-analysis of clinical trial data had reported an increased relative risk of atrial fibrillation in patients receiving Ivabradine.

7. Andres Ricardo Perez Riera1, *et al.,* has done research work entitled -Ivabradine: Just another new pharmacological option for Heart rate control. According to the study, Clinical indications of Ivabradine was found to be safe and effective for symptomatic management of chronic stable angina with HR $\geq$ 60-70bpm, congestive heart failure, inappropriate sinus tachycardia(ST) and postural orthostatic tachycardia syndrome (POTS).

8. Pietroscicchitano1, *et al.,* has done research work entitled -Ivabradine, coronary artery disease, and heart failure: beyond rhythm control. According to the study, Ivabradine was a fundamental molecule for the treatment of Coronary artery disease patients with or without left ventricular dysfunction. Ivabradine effectively controls heart rate both at rest and during exercise and improves the symptoms of angina and the exercise capacity of patients with chronic stable angina.

9. Juan Carlos Kaski1, *et al.,* has done research work entitled - Role of Ivabradine in management of stable angina in patients with different clinical profiles. According to the study; in angina patients with Congestive heart failure and Left ventricular systolic dysfunction, the use of Ivabradine reduced recurrent hospitalization, improved prognosis and quality of life. Ivabradine a drug that selectively reduces heart rate for the treatment of stable angina pectoris in different clinical conditions, including patients with preserved or impaired Left ventricular function by reducing heart rate without affecting myocardial inotropic function or coronary vasomotor tone.

10. Salem M, *et al.,* have done research work entitled -Safety and efficacy of Ivabradine in patients with acute ST segment elevation MI. According to the study, the aim was to explore safety and efficacy of Ivabradine in patients with STEMI associated with left ventricular dysfunction. 100 patients received 5mg Ivabradine twice a day in addition to the conventional treatment, while 100 patients received the conventional only. Ivabradine didn't show a significant impact on major adverse cardiac events when added to conventional treatment. The study reported that adding Ivabradine to the conventional therapy for patients presented with anterior STEMI was successfully repufused did not improve clinical outcomes.

11. Sathyamurthy I, *et al.,* have done research work entitled - Ivabradine: Evidence and current role in cardiovascular diseases and other emerging indications. According to the study, Ivabradine provides additional benefits when used in combination with the other antianginal drugs such as beta-blockers (except diltiazem and verapamil). In symptomatic patients, despite treatment with beta-blockers, adding ivabradine provides a significant benefit.

12. Roberto Ferrari, *et al.,* has done research work entitled -Ivabradine in the management of coronary artery disease with or without left ventricular dysfunction or heart failure. According to the study, Provided that it is used at the recommended dose, Ivabradine has been proved to be an effective, useful antianginal agent in patients with preserved or reduced EF.

Ivabradine improved the condition of patient, but did not worsen the prognosis. Thus, it seems to be a logical to use Ivabradine when elevated heart rate remains high.

13. Marina Pascual Izco, *et al.,* has done research work entitled -Ivabradine in acute heart failure: Effects on heart rate and hemodynamic parameters in a randomized and controlled swine trial. According to the study, Ivabradine reduces heart rate and increases stroke volume without modifying systemic or left filling pressures in a swine model of acute heart failure. Future studies with specific heart rate targets are needed.

14. Michael B€ohm, *et al.,* have done research work entitled – Effect of Visit-to-Visit Variation of Heart Rate and Systolic Blood Pressure on Outcomes in Chronic Systolic Heart Failure: Results From the Systolic Heart Failure Treatment With the If Inhibitor Ivabradine Trial (SHIFT) Trial. According to the study, beyond high HR and low SBP, low HR-CV and low SBP-CV are predictors of cardiovascular outcomes with additive effects on risk in HF, but with an unknown effect size. Beyond HR reduction, Ivabradine increases HR-CV. Low visit-to-visit variation of HR and SBP might signal risk of cardiovascular outcomes in systolic HF.

15. Riccardo Cappato, *et al.,* has done research work entitled – Clinical Efficacy of Ivabradine in Patients with Inappropriate Sinus Tachycardia. According to the study, ivabradine significantly improved symptoms associated with inappropriate sinus tachycardia and completely eliminated them in approximately half of the patients. Ivabradine administration was also associated with a significant increase in exercise performance. No cardiovascular side effects were observed in any patients while taking ivabradine.

16. Kim Fox, *et al.,* has done research work entitled – Effect of Ivabradine in patients with left-ventricular systolic dysfunction: a pooled analysis of individual patient data from the BEAUTIFUL and SHIFT trials. According to the study, Ivabradine may be important for the improvement of clinical outcomes in patients with LV systolic dysfunction and heart rate ≥70 b.p.m., whatever the primary clinical presentation (CAD or HF) or clinical status (NYHA class).

17. Sally Tse, *et al.,* has done research work entitled – Ivabradine (Corlanor) for Heart Failure: The First Selective and Specific Inhibitor. According to the study, Ivabradine is the only agent shown to clinically lower the heart rate without negative inotropism or effects on conduction and contractility. As a result, ivabradine is not associated with the adverse events.

18. Jean-Claude Tardif, *et al.,* has done research work entitled – Ivabradine in clinical practice: benefits of *If* inhibition. According to the study, Ivabradine is a selective and specific If inhibitor with anti-anginal and anti-ischaemic effects that have been shown to be non-inferior to those of the b-blocker atenolol and the calcium channel blocker amlodipine. Unlike b-blockers, Ivabradine is devoid of intrinsic negative inotropic effects and does not affect coronary vasomotion. A whole range of patients with angina may benefit from exclusive heart rate reduction with Ivabradine, including those with contraindications or intolerance to the use of b-

blockers and patients who are insufficiently controlled by b-blockers or calcium-channel blockers.

19. Ahmed Ashraf Eissa, *et al.,* has done research work entitled – Effect of Ivabradine on the Infarct Size and Remodeling in Patients with ST Elevation Myocardial Infarction. According to the study, In the setting of STEMI treated with Primary Percutaneous Coronary Intervention (PPCI), Ivabradine significantly reduced the HR. In a subgroup of Diabetic patients with HR>100 bpm, Ivabradine significantly reduced the echocardiographic LVESD and improve the SPECT LVEF.

20. Ivano Bonadei, *et al.,* has done research work entitled – Is There a Role for Ivabradine Beyond its Conventional Use?. According to the study, despite its indication in chronic HF and stable angina patients, ivabradine may be safe and effective in several other clinical settings.

21. Leonardo Calò, *et al.,* has done research work entitled – Is Efficacy of Ivabradine administration in patients affected by inappropriate sinus tachycardia. According to the study, Ivabradine could represent an effective and safe alternative to calcium channel blockers and beta-blockers for treatment of inappropriate sinus tachycardia (IST) patients (especially in those with higher initial HR), leading to symptom relief and determining a progressive increase of stress tolerance.

Chapter-4

Need for the Study

- ➢ Ivabradine is prescribed in patients with Acute Myocardial Infarction to decrease the Heart Rates (HR) without any effect on hemodynamic or myocardial contractility.
- ➢ It is prescribed in case of Heart Rate $\geq$ 70bpm.
- ➢ This study will show the safety and adverse effects of Ivabradine in patients with Acute Myocardial Infarction.

This study will improve patient safety and quality of life

Study on Safety and Adverse Effects of Ivabradine in Acute Myocardial Infarction

Chapter-4

Aims and Objectives

Aim of the Study

To evaluate the safety and adverse effects of Ivabradine in patients with Acute Myocardial Infarction

Objectives

Primary objective: To evaluate safety of Ivabradine in patients with acute myocardial infarction:

Clinical symptoms:
- ➤ Blood Pressure
- ➤ ECG
- ➤ Heart Rate
- ➤ T wave inversion

Secondary objective: To evaluate Adverse effects of Ivabradine in patients with Acute Myocardial Infarction.

Chapter-5

Plan of the Study

- ➢ Literature survey.
- ➢ Collection of cases on Ivabradine in Acute Myocardial Infarction in tertiary Care Hospital according to the designed proforma.
- ➢ To analyse the safety and adverse effects of Ivabradine in Acute Myocardial Infarction.

Statistical analysis and reporting for better patient care.

Study on Safety and Adverse Effects of Ivabradine in Acute Myocardial Infarction

Chapter-6
Methodology

Study Site:

The study will be carried out at the Care Hospital, Banjara Hills. The data required was collected from the cardiology department. During the study period, the data were collected on Ivabradine in Acute Myocardial Infarction in the hospital and on follow up visits as well.

Study Design:

Retrospective Observational study: A retrospective study is a study in which a search is made for a relationship between one phenomenon (usually present) and another that occurred in the past. The advantages of retrospective study are its small scale, usually short time for achievement and its application to rare diseases. Its drawbacks are certain important statistics cannot be deliberate and large biases may be introduced in the recall of past exposure to risk factors.

Study Duration:

6 months

Study Criteria:

S.No	Inclusion Criteria :	Exclusion Criteria:
1	Patients of all age groups.	Patients with Heart rate <100bpm is excluded
2	Sinus rhythm (tachycardia)>100bpm.	Patients who are not willing to participate.

Study Procedure:

1. During the study period, the data were collected on daily basis, as Ivabradine in Acute Myocardial Infarction in the hospital and on follow - up visits as well.
2. The collected data were entered in data collection forms designed for the recording of only those parameters essential to establish the objectives of the study.
3. The data collection form used for the study was designed on Google drive, after which, it was entered month wise into excel.
4. The results were then obtained after filtering for the required data, figures and percentages.

Sources of Data:

All the necessary and relevant information was collected from the Cardiology department in tertiary care hospital.

Statistical Analysis:

Continuous variables are represented as mean and standard deviation where data follows normal distribution, otherwise as median with range. Categorical variables are represented as frequencies and percentages. Data was analyzed using R studio.

Chapter-7

Results

It is a Retrospective and observational study in which we identified and evaluated the Safety and Adverse effects of Ivabradine in Patients with Acute Myocardial Infarction. A total number of Ivabradine cases in Acute Myocardial Infarction (N = 65) are enlisted in the study, during the study period, the data of patients were categorized according to age - groups, gender, diagnosis, co-morbidities and adverse effects. The percentages were calculated and established.

Table No- 3: Percentage of Ivabradine in Acute Myocardial Infarction According to Gender:

Gender	No. of Patients	Percentage (%)
Male	45	69.23
Female	20	30.77

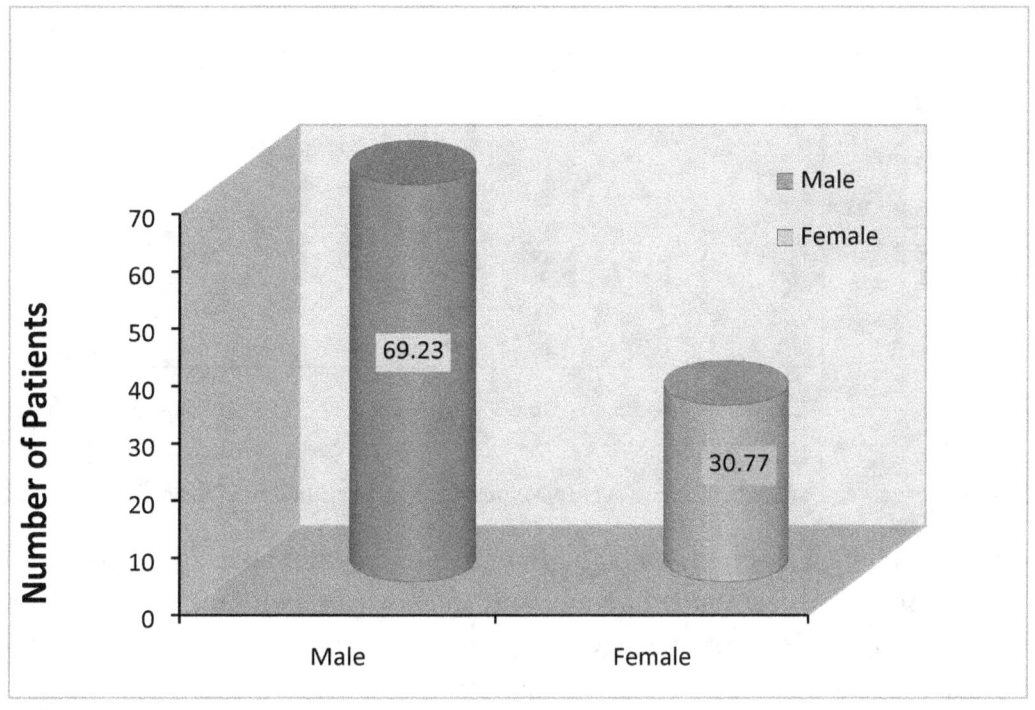

Figure No- 17: Percentage of Ivabradine in Acute Myocardial Infarction According to Gender

According to the gender, Ivabradine in Acute Myocardial Infarction cases, the male population which was 45(69.23) in number was significantly more than the female population which was 20(30.77) in number which is summarized in **(Table 3 and Figure 17).**

Table No- 4: Percentage of Ivabradine in Acute Myocardial Infarction Cases according to Age Groups

Age Group	No. of Patients	Percentage (%)
25-35	1	1.54
36-45	3	4.62
46-55	16	24.62
56-65	16	24.62
66-75	20	30.77
76-85	9	13.85

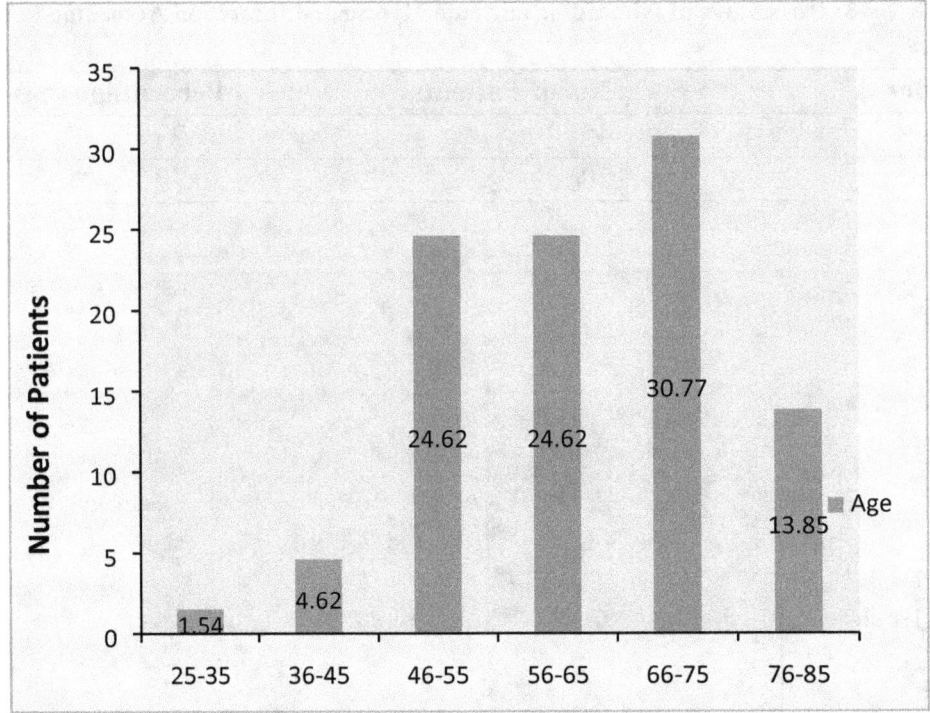

Figure No- 18: Percentage of Ivabradine in Acute Myocardial Infarction Cases according to Age Groups

In our study, according to the age group, the highest number of patients who are on Ivabradine in Acute Myocardial Infarction were found to be within the age group of 66–75(N=20). The graph rising from the age group 25-35years and again falling to an age group of 76-85years **(shown in Table- 4 and Figure- 18).**

Table No- 5: Percentage of Ivabradine in Acute Myocardial Infarction Cases according to Diagnosis

Diagnosis	No. of Patients	Percentage (%)
AWMI(Anterior Wall Myocardial Infarction)	44	67.69
IWMI(Inferior Wall Myocardial Infarction)	2	3.08
PWMI(Posterior Wall Myocardial Infarction)	2	3.08
LWMI(Lateral Wall Myocardial Infarction)	1	1.54
NSTEMI(Non ST-elevated Myocardial Infarction)	11	16.92
STEMI(ST-elevated Myocardial Infarction)	2	3.08
Sev.LV dysfunction	3	4.62

In our study out of 65 patients, the diagnosis was found to be AWMI with percentage of 67.69(N=44), IWMI with percentage of 3.08(N=2), PWMI with percentage of 3.08(N=2), LWMI with percentage of 1.54(N=1), NSTEMI with percentage of 16.92(N=11), STEMI with percentage of 3.08(N=2) and Sev.LV Dysfunction with percentage of 4.62(N=3) **(Shown in Table 5 and Figure 19).**

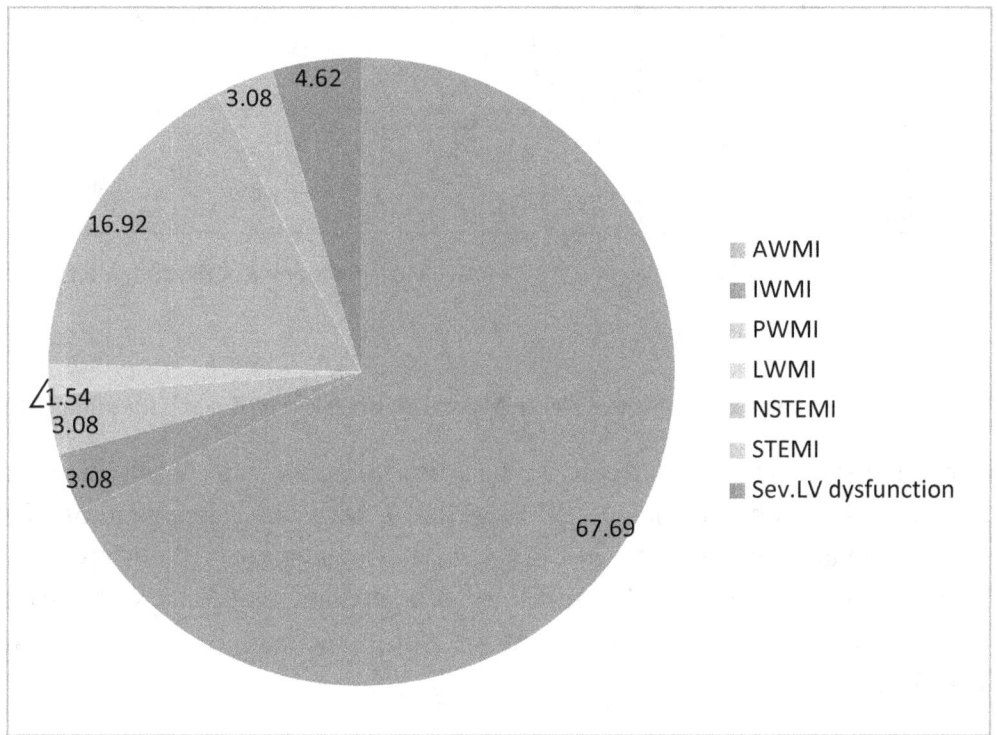

Figure No- 19: Percentage of Ivabradine in Acute Myocardial Infarction Cases According To Diagnosis

Table No- 6: Percentage of Ivabradine in Acute Myocardial Infarction cases with other Co-morbidities

Co-morbidities	No. of Patients	Percentage (%)
HTN(Hypertension)	37	30.33
DM(Diabetes Mellitus)	34	27.87
CAD(Coronary Artery Disease)	24	19.67
ACS(Acute Coronary Syndrome)	13	10.66
Sev.LV dysfunction	4	3.28
Anaemia	2	1.64
CKD(Chronic Kidney Disease)	5	4.10
Hypothyroidism	3	2.46

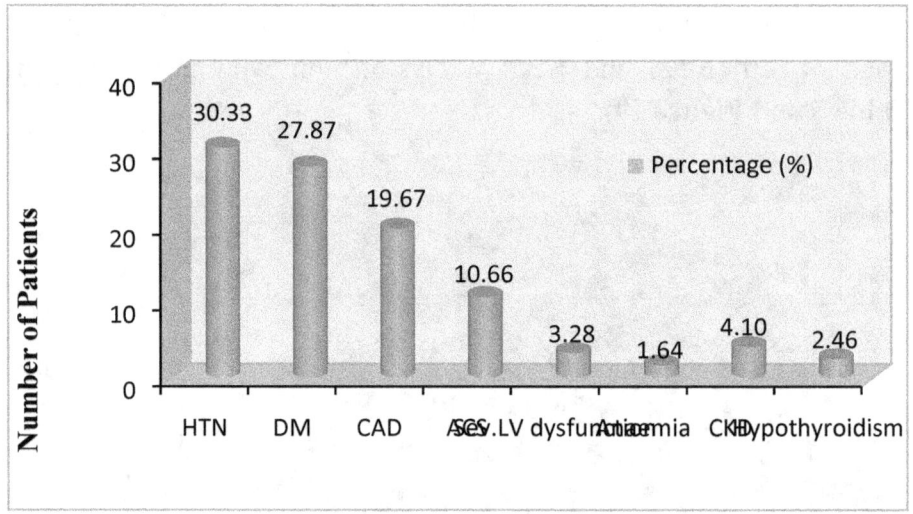

Figure No- 20: Percentage of Ivabradine in Acute Myocardial Infarction Cases with other Co-morbidities

In our study, the patients of Acute Myocardial Infarction with Ivabradine had the Co-morbidities of HTN with percentage of 30.33(N=37), DM with percentage of 27.87(N=34), CAD with percentage of 19.67(N=24), ACS with percentage of 10.66(N=13), CKD with percentage of 4.10(N=5), Sev.LV Dysfunction with percentage of 3.28(N=4), Anaemia with percentage of 1.64(N=2) and Hypothyroidism with percentage of 2.46(N=30) **(Shown in Table-6 and Figure- 20).**

Table No- 7: Adverse Effects in Acute Myocardial Infarction Patients due to Ivabradine

Adverse Effects	No. of Patients	Percentage (%)
Atrial Fibrillation	1	1.54
Bradycardia	3	4.62
Without Adverse effects	61	93.84

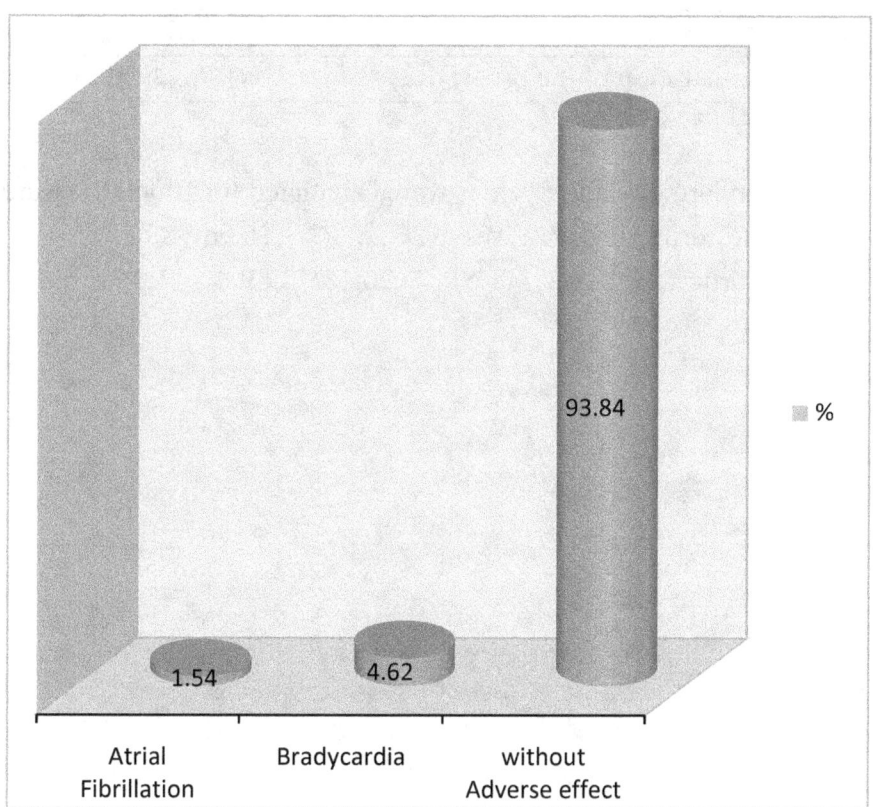

Figure No-21: Adverse Effects in Acute Myocardial Infarction Patients due to Ivabradine

In our study total N=65 patients 3 patients had the adverse effects of Bradycardia with percentage of 4.62 and 1 patient had an episode of Atrial Fibrillation with percentage of 1.54**(Shown in Table 7 and Figure 21).**

The Mean Standard deviation values were calculated for Heart rates (HR), before administration of the drug – HR1 was 105.26 $\pm$ 13.05 and after administration the drug – HR2 was 81.95 $\pm$ 12.03 respectively.

Table No- 8: Mean $\pm$ Standard Deviation for Heart Rates

S.No	Parameters – Blood Pressure	Mean $\pm$ Standard deviation
1	Before administration of the drug(BP1)	126.28 $\pm$ 26.96/78.77 $\pm$ 14.20
2	After administration of the drug(BP2)	111.51 $\pm$ 20.18/72.89 $\pm$ 11.46

Table No- 9: Mean $\pm$ Standard Deviation for Blood Pressure

S.No	Parameters – Heart rates	Mean $\pm$ Standard Deviation
1	Before administration of the drug(HR1)	105.26 $\pm$ 13.05
2	After administration of the drug(HR2)	81.95 $\pm$ 12.03

The Mean Standard deviation values were calculated for Blood Pressures (BP), before administration of the drug – BP1 was 126.28 $\pm$ 26.96/78.77 $\pm$ 14.20 and after administration of the drug – BP2 was 111.51 $\pm$ 20.18/72.89 $\pm$ 11.46 respectively.

Chapter-8

Discussion

Heart Rate (HR) is a precisely regulated variable, which plays a critical role in health and disease. Elevated resting heart rate is a significant predictor of all cause and cardiovascular mortality in the general population and patients with cardiovascular disease (CVD). Coronary Artery Disease (CAD) represents the most common cause of death in middle aged and older adults in European Countries. Acute myocardial infarction most commonly occurs when thrombus formation results in complete occlusion of a major epicardial coronary vessel.

Ivabradine is a heart rate lowering agent that acts by selectively and specifically inhibiting the cardiac pacemaker current I_f a mixed sodium-potassium inward current that controls the spontaneous diastolic depolarization in the Sino atrial (SA) node and hence regulates the heart rate.

Ivabradine should currently used as a second-line agent for managing angina or as first-line treatment if the patient is intolerated to beta-blockers or if there are contraindications. Ivabradine may be administered early to patients with successful PCI (Primary precutaneous intervention) for anterior STEMI with an impaired left ventricular function and high HR and sinus rhythm.

This study is a retrospective and observational study in which we identified and evaluated the safety and adverse effects of Ivabradine in patients with Acute Myocardial Infarction. Ivabradine is prescribed in patients with HR of $\geq$ 70bpm. The drug was prescribed with the dose of 2.5mg BID, if the heart rate does not decreases then the dose was increased to 5mg BID. The other studies also shown the reduction of heart rate with the dose of 2.5mg BID, 5mg BID and 7.5mg BID respectively.

A total number of Ivabradine cases in Acute Myocardial Infarction (N=65) are enrolled in the study, during the study period. The data of patients were categorized according to age groups, gender, diagnosis, co-morbidities and adverse effects.

According to the gender the male population percentage was 69.23% (N=45) and female population percentage was 30.77% (N=20). In this study, male population was more predominant than the female population. Other studies also show that the male population percentage was more when compared to the female population percentage. According to the age groups 66-75years (N=20) patients are more prevalent to Acute Myocardial Infarction. Many studies have shown that more than 50% of the people older than 60years have significant CAD. According to the study the diagnosis was more prevalent on Anterior wall myocardial infarction (AWMI) with percentage of 67.69% (N=44) when compared to others.

Discussion

Out of N= 65 patients, the adverse effects was found to be N=4 with the percentage of 6.16%, whereas Atrial Fibrillation percentage was 1.54% (N=1) and Bradycardia with percentage of 4.62% (N= 3). Many studies show that the most common adverse reactions occurring in $\geq 1\%$ of patients who received Ivabradine included bradycardia (10%) and atrial fibrillation (8.3%). The study shows the Safety and Adverse effects of Ivabradine in patients with Acute Myocardial Infarction in tertiary Care Hospital.

Other studies clearly demonstrated that Ivabradine was associated with a 42% reduction in hospitalization for MI. In patients with heart rates ≥ 70bpm, there was 73% reduction in hospitalization for MI and 59% reduction in Coronary revascularization.

Chapter-8
Conclusion

- ➢ Ivabradine is a drug that reduces heart rate by inhibiting the cardiac I_f channels in their open state. Ivabradine should only be prescribed to patients who are already on the maximal-tolerated beta-blocker dose with a heart rate that remains 70bpm or higher.
- ➢ Ivabradine is the only agent shown to clinically lower the heart rate without negative inotropism or effects on conduction and contractility. As a result, ivabradine is not associated with the adverse events typically encountered with other bradycardic agents.
- ➢ Our study shows that, Ivabradine was safe for reduction of heart rate in acute myocardial infarction.

Conclusion

Study on Safety and Adverse Effects of Ivabradine in Acute Myocardial Infarction

Chapter-10
Summary

Myocardial Infarction (MI) is the term used for an event of heart attack which is due to blockage of the arteries resulting in absent of blood flow to the myocardial and causing death of the muscle called as infarction. Acute myocardial infarction has been divided into ST Elevation and Non ST- Elevation Myocardial Infarction.

Acute coronary syndromes lead to myocyte injury and subsequent death. In the clinical setting, their classification is based upon electrocardiographic presentation; ST-segment elevation myocardial infarction (STEMI; ≥ 2 mm ST segment elevation and prominent T waves on the electrocardiogram) and Non-ST-segment elevation myocardial infarction (NSTEMI; symptoms of acute coronary syndrome exemplified by ST-segment depression and T wave inversion on the electrocardiogram).

ST- Elevated Myocardial Infarction has been classified into six types. They are Anterior ST Segment Elevation MI, Inferior ST Segment Elevation MI, Posterior ST Segment Elevation MI, Anterolateral ST Segment Elevation MI, Acute MI with a Right Bundle Branch Block, Acute MI with a Left Bundle Branch Block.

Early treatment aims to reduce the extent of myocardial damage as the myocardium is damaged by a diminished oxygen supply due to the obstructed coronary artery.

Ivabradine is a Hyperpolarization-activated Cyclic Nucleotide-gated Channel Blocker. The mechanism of action of ivabradine is as a Hyperpolarization-activated Cyclic Nucleotide-gated Channel Antagonist. Ivabradine is a novel, specific HR-lowering agent that acts in SAN cells by selectively inhibiting the pacemaker If current in a dose-dependent manner.

The most common adverse events associated with Ivabradine were Bradycardia, Atrial Fibrillation, Hypertension, Luminous Phenomena.

It is a Retrospective and Observational study in which we identified and evaluated the Safety and Adverse effects of Ivabradine in patients with Acute Myocardial Infarction. The male population was more predominant than the female population. In the age groups 66-75years patients are more prevalent to acute myocardial infarction. In this study the diagnosis was more prevalent on AWMI with percentage of 67.69% (N=44), when compared to others. Out of N=65 patients, the adverse effects was found to be N=4 with the percentage of 6.16%, where as Atrial Fibrillation with the percentage of 1.54% (N=1) and Bradycardia with percentage of 4.62% (N=3).

Summary

This study shows the safety and adverse effects of Ivabradine in patients with Acute Myocardial Infarction in Tertiary Care Hospital.

Chapter-11
Future Scope and Limitations of the Study

Future Scope of the Study

- ➢ Parallel group randomization study.
- ➢ Prospective follow up study.
- ➢ To increase the study population.

To carry forward the study for better patient care

Limitations of the Study

- ➢ The results of this study may not be completely generalized as the sample size was restricted.
- ➢ The study was carried out in care hospitals, Banjara Hills which is a tertiary Care hospital and hence the cases we study may not be representative of those occurring in the general population.
- ➢ Selective bias.
- ➢ Retrospective study rather than a prospective study.
- ➢ It was not a multi centered study to arrive at a concrete conclusion.

Study on Safety and Adverse Effects of Ivabradine in Acute Myocardial Infarction

Bibliography

1. Mayo Clinic et al., Heart attack. National Heart, Lung, and Blood Institute. 2018; 1

2. Koushik Reddy, Asma Khaliq, Robert J Henning et al., Recent advances in the diagnosis and treatment of acute myocardial infarction; World J Cardiol 2015 May 26;Vol- 7(5): 243-276 ISSN 1949-8462.

3. Kristian Thygesen et al., European Heart Journal, Volume 40, Issue 3, 14 January 2019, 237– 269.

4. Andrew R Chapman, Philip D Adamson, Nicholas L Mills et al., Assessment and classification of patients with myocardial injury and infarction in clinical practice. Heart2017;103:10–18

5. Joshua Chadwick jayaraj, et al., Epidemiology of Myocardial Infarction.Nov-5th 2018.

6. Philip Aylward et al., Acute Myocardial Infarction: early treatment.Aust Prescr 1996;19:52-41.

7. Richard N. Fogoros, Md et al., Causes of Myocardial Infarction (Heart Attack).October 05, 2018.

8. Medically reviewed by Debra Sullivan et al., November 30, 2017; Written by Brindles Lee Macon, Winnie Yu, and Lauren Reed-Guy. Acute Myocardial Infarction.

9. Br J Gen Pract et al., Signs and symptoms in diagnosing acute myocardial infarction and acute coronary syndrome: a diagnostic meta-analysis. 2008 Feb 1; 58(547): e1–e8.

10. Rudradev Pandey, et al., Diagnosis of acute myocardial infarction. The journal of the Association of Physicians of India 59 Suppl(12):8-13 · December 2011 with 54 Reads.

11. Vedika Rathore, et al., Risk factors for Acute Myocardial Infarction.EJMI 2018;2(1):17.

12. Sulfi S and Adam Timmis et al., Ivabradine – the first selective sinus node If channel inhibitor in the treatment of stable angina. International Journal of Clinical Practice 2006; 60(2): 222– 228.

13. Susan Wilansky, et al., Complications of myocardial infarction.Critical Care Medicine. 2007;35(8):S348-S354.

14. Sandhya S, Mohanraj P. et al., Clinical presentation, risk factors, complications and outcome of acute myocardial infarction in elderly patients.Int J Res Med Sci. 2017 Nov;5(11):4765-4769.

15. Claudio Borghi & Ettore Ambrosioni et al., Primary and Secondary Prevention of Myocardial Infarction. Clin. and Exper, Hypertension, 18(3&4), 547-558 (1996).

16. Richard E. Collins, MD et al., After MI, lifestyle modifications play key role in prevention. GISSI-Prevenzione Investigators. Lancet. 1999;354:447-455. Lichtenstein A. Circulation. 2006;114:82-96.

17. Dr. Maxwell S et al., Emergency management of acute myocardial infarction. British Journal of Clinical Pharmacology; 1999; 48(3): 284–298.

18. By Richard N. Fogoros, MD et al., Non-ST Segment Myocardial Infarction. December 19, 2018.

19. Kingma Jr, J.G et al., (2018) Myocardial Infarction: An Overview of STEMI and NSTEMI

20. Lal C Daga, Upendra Kaul, Aijaz Mansoor. Approach to STEMI and NSTEMI.Supplement to JAPI. December 2011. VOL.59.

21. Dr Araz Rawshani et al., NSTEMI (Non ST Elevation Myocardial Infarction) & Unstable Angina: Diagnosis, Criteria, ECG, Management.First published: 2016-09-03.

22. Steven Lome et al., MI ECG Patterns you must know. Thygesen K, et al. Third Universal Definition of Myocardial Infarction. Circulation. 2012;doi: 10.1161/CIR. 13(3)1826-1058.

23. Chou's et al., Electrocardiography in Clinical Practice: Adult and Pediatric, 6e.Surawicz B, et al. AHA/ACCF/HRS Recommendations for the Standardization and Interpretation of the Electrocardiogram. Circulation. 2009;Anterior Wall ST Segment Elevation MI ECG Review.

24. Morton Kern, MD et al., The Different Presentations of Acute STEMI: What Problems Should the Cath Lab Look For. Volume 18 - Issue 11 - November 2010.

25. Bryan Wilner, MD, et al., LBBB in Patients With Suspected MI: An Evolving Paradigm. Feb 28, 2017. 26. Riccioni et al., Ivabradine: from molecular basis to clinical effectiveness. 2010 Mar;27(3):160-7.

26. National Center for Biotechnology Information. PubChem Compound Database; CID=132999. Ivabradine.accessed Mar. 14, 2019.

27. Jacob S. Koruth, et al., The Clinical Use of Ivabradine. JACC VOL. 70, NO. 14, 2017 OCTOBER 3, 2017:1777–84.

28. Sarah E. Petite, et al., Role of the Funny Current Inhibitor Ivabradine in Cardiac Pharmacotherapy: A Systematic Review. American Journal of Therapeutics 0, 1–20 (2016).

29. Gian Marco Rosa, et al., An evaluation of the pharmacokinetics and pharmacodynamics of ivabradine for the treatment of heart failure. (2014) 10(2):279-291.

30. Sally Tse, et al., Ivabradine (Corlanor) for Heart Failure: The First Selective and Specific If Inhibitor. P T. 2015 Dec; 40(12): 810–814.

31. Thamir M. Alshammari, et al., Ivabradine: Do the Benefits Outweigh the Risks? Journal of Cardiovascular Pharmacology and Therapeutics 1-9. The Author(s) 2016.

32. Ivabradine. U.S. National Library of Medicine 8600 Rockville Pike, Bethesda, MD .20894.U.S.DepartmentofHealthandHumanServices NationalInstitutes of Health Page last updated: 28 January 2019.

33. Barry Greenberg, Amrit Dosanjh et al., An evaluation of the pharmacokinetic and pharmacodynamics of ivabradine for the treatment of heart failure. Cariology advisor. Ivabradine. 2014; 10: 279-291.

34. DiFrancesco D et al., The contribution of pacemaker current (if) to generation of spontaneous activity in rabbit sino-atrial node myocytes. J. Physiol. 1991; 34: 23–40.

35. Brown H.F, DiFrancesco D. Noble S.J et al., How does adrenaline accelerate the heart? Nature1979; 280: 235–236.

36. Koester R, Kaehler J. Meinertz T et al., Ivabradine for the treatment of stable angina pectoris in octogenarians. Clin. Res. Cardiol. 2011; 100: 121–128.

37. Camm, A.J, Lau C.P et al., Electrophysiological effects of a single intravenous administration of ivabradine in adult patients with normal baseline electrophysiology. Drugs R. D. 2003; 4: 83–89.

38. Heusch G, Skyschally A, Gres P, Van Caster P, Schilawa D, Schulz, R. Improvement of regional myocardial blood flow and function and reduction of infarct size with ivabradine: Protection beyond heart rate reduction. Eur. Heart J. 2008; 29(18): 2265–2275.

39. Sathyamurthy a, Sanket Newale et al., Ivabradine: Evidence and current role in cardiovascular diseases and other emerging indications. Indian Heart Journal 70 (2018) S435 - S441.

40. Difrancesco D, Camm JA et al., Heart rate lowering by specific and selective I(f) current inhibition with ivabradine: a new therapeutic perspective in cardiovascular disease. 2004; 64(16): 1757-65.

41. Guglin M, et al., Heart rate reduction in heart failure: ivabradine or beta blockers? 2013 Jul; vol-18(4): page no-517-28. doi: 10.1007/s10741-012-9347-6.

42. Foster JL, Bobadilla RV et al., Ivabradine, a novel medication for treatment of heart failure with reduced ejection fraction. 2016 Nov; 28(11): 576-582. doi: 10.1002/2327-6924.12371.

43. Steg P, Lopez-de-Sà E et al., Safety of intravenous ivabradine in acute ST-segment elevation myocardial infarction patients treated with primary percutaneous coronary intervention: a randomized, placebo-controlled, double-blind, pilot study. 2013 Sep;2(3):270-9. doi: 10.1177/2048872613489305.

44. Irmina Urbanek, Krzysztof Kaczmarek et al., Risk-benefit assessment of ivabradine in the treatment of chronic heart failure; 2014; 6: 47–54.

45. Jules C. Hancox, Dario Melgari, et al., hERG potassium channel inhibition by ivabradine may contribute to QT prolongation and risk of torsades de pointes: 2015 Aug; 6(4): 177–179.

46. Andres Ricardo Perez Riera, Luiz Carlos de Abreu et al., Ivabradine: Just another New Pharmacological Option for Heart Rate Control? S4:001. doi:10.4172/2155-9880.S4-001 ISSN: 2155-9880.

47. Scicchitano P, Cortese F, et al., Ivabradine, coronary artery disease, and heart failure: beyond rhythm control. 2014 Jun 3; 8: pg no: 689-700.

48. Kaski JC, Gloekler S et al., Role of ivabradine in management of stable angina in patients with different clinical profiles. 2018 Mar 9; 5(1): e000725. doi: 10.1136/openhrt-2017-000725.

49. Salem, Mohamed Salah, et al., Safety and efficacy of Ivabradine in patients with acute STsegment elevation myocardial infarction. 2015; Vol. 2(1), pp. 007-011, ISSN: 2146-3133.

50. Sathyamurthy, Newale et al., Ivabradine: Evidence and current role in cardiovascular diseases and other emerging indications; Indian Heart Journal 70 (2018) S435-S441

51. Roberto Ferrari et al., Ivabradine in the management of coronary artery disease with or with

52. out left ventricular dysfunction or heart failure; European Heart Journal Supplements (2015) 17 (Supplement G), G24–G29.

53. Pascual Izco, Ramírez-Carracedo et al., Ivabradine in acute heart failure: Effects on heart rate and hemodynamic parameters in a randomized and controlled swine trial. 2018 Aug 29. doi: 10.5603/CJ.a2018.0078.

54. Michael B€ohm, MD; Michele Robertson et al., Effect of Visit-to-Visit Variation of Heart Rate and Systolic Blood Pressure on Outcomes in Chronic Systolic Heart Failure: Results From the Systolic Heart Failure Treatment With the If Inhibitor Ivabradine Trial (SHIFT) Trial; American Heart Assoc. 2016;5: e002160 doi: 10.1161/JAHA.115.002160.

55. Riccardo Cappato, Serenella Castelvecchio et al., Clinical Efficacy of Ivabradine in Patients With Inappropriate Sinus Tachycardia A Prospective, Randomized, Placebo-Controlled, Double-Blind, Crossover Evaluation; American College of Cardiology Foundation, 2012; ISSN 0735-1097 Vol. 60, No. 15.

56. Fox K, Komajda M et al., Effect of ivabradine in patients with left-ventricular systolic dysfunction: a pooled analysis of individual patient data from the BEAUTIFUL and SHIFT trials; 2013 Aug;34(29):2263-70.

57. Sally Tse, and Nissa Mazzola et al., Ivabradine (Corlanor) for Heart Failure: The First Selective and Specific If Inhibitor; 2015 Dec; 40(12): 810–814.

58. Jean-Claude Tardif et al., Ivabradine in clinical practice: benefits of If inhibition European Heart Journal Supplements, 2005; Volume 7, Issue suppl H, 1 September 2005, Pages 29-32.

59. Ahmed Ashraf Eissa et al., Effect of Ivabradine on the Infarct Size and Remodeling in Patients with ST Elevation Myocardial Infarction; 2018; Volume 71, Issue 16 Supplement.

60. Bonadei, Vizzardi et al., Is there a role for ivabradine beyond its conventional use? 2014 Aug; 32(4): 189-92.

61. Calò, Rebecchi et al., Efficacy of ivabradine administration in patients affected by inappropriate sinus tachycardia; 2010 Sep;7(9):1318-23.

Study Proforma

Table No- 1: Patient Demographic Details

Patient Id No:	Age:	Gender:
Weight:	DOA:	DOD:
Current Complaints:		

Table No- 2: Patient History

Medical History:				
Medication History:				
Drug	**Dose**	**Route**	**Frequency**	**Duration**
Social History:				
Smoking: Yes/No If yes _____packs/day Alcohol: Yes/No If yes _____ml/day Chewing tobacco :Yes/No If yes _____quantity				
Allergies:				
Family History:				
Surgical History				

Table No- 3: Vital Signs

Temperature :   Heart rate :   Blood pressure :

Table No- 4: Lab Findings

ECG:
2D ECHO: RWMA+: LAD: ▇ RCA: ▇ LCX: ▇
LVEF: Improved: ▇ Worsened: ▇

Table No- 5: Diagnosis

AWMI: ▇	IWMI: ▇	PWMI: ▇
NSTEMI: ▇	INFERIOR AND POSTERIOR MI: ▇	
NSTEMI: ▇	Others: ▇	

Table No- 6: Current Therapy

Drug	Dose	Route	Frequency	Duration					
				D1	D2	D3	D4	D5	D6

Discharge Dose

Other prescribed drugs in combination with Ivabradine:

Table No- 7: Adverse Effects

Atrial Fibrillation:	■	**Luminous Phenomena:**	■
Bradycardia:	■	**Hypertension:**	■
Others:			

Table No 8: Follow up

Chief Components on Presentations :
ECG: if available **New Changes:**
2D ECHO:
ANGIOGRAM: if available

Inform Consent Waiver Letter

To
The Chairman,
Institutional Ethics Committee,
Care Convergence Centre,
Road No. 10, Banjara Hills,
Hyderabad 500034, India.

<u>Reference:</u> **Title: "Study on safety and adverse effects of Ivabradine in patients with acute myocardial infarction in tertiary Care Hospital."**

Sub: Request for Inform Consent waiver

Respected Sir,

I request you to allow waiver for consent from study patients as this study entails only a retrospective chart review. The procedure that is being studied is part of routine clinical management of patients and not experimental. All procedures were performed only after obtaining written informed consent from patients according to hospital protocol. The study does not involve collection of data related to HIV/AIDS, genetic information, mental health information and substance abuse information or any other that would compromise the dignity or safety of the patient.

The data collected will be stored as a password protected soft copy and will be accessible only to the Guide/ investigator.

Thanking You,

Yours truly,

Chirumamilla Murali krishna, Thanneru Keerthi,
Marupakala Shirisha.

Ethical Declaration

CARE Hospital
Institutional Ethics Committee
Regn. No. ECR/94/Inst./AP/2013/RR-16

Date: 11th Dec 2018

To

M. Sirisha, T. Keerthi and Ch. Murali Krishna
Pharma-D Students,
Dept. of Cardiology,
Guru Nanak Institutions Technical Campus-School of Pharmacy,
Ibrahimpatnam, R R District, Telangana

Dear Sir/Madam,

Ref: Thesis Protocol - "Study on safety and adverse effects of Ivabradine in patients with acute myocardial infarction in tertiary care hospital"

Sub: EC approval of the A5 of Institutional Ethics Committee Meeting held on 2nd Dec 2018

The Institutional Ethics Committee met on Sunday, 2nd Dec 2018 at 10.00 A.M. at Board Room (PACE), Care Convergence Centre, 8-2-595/2/B, Road No.10, Banjara Hills, Hyderabad–500034 for the monthly meeting and reviewed/discussed the thesis protocol submitted by you with the following documents of above mentioned study to be carried out at the site Dept. of Cardiology, Care Hospital, Hyderabad.

1. Letter from the student to the Chairman, Institutional Ethics Committee dated 15th Nov 2018
2. Protocol
3. Waiver of Informed Consent

The following Members of Institutional Ethics Committee are present for the study

Name	Qualification	Designation	Field
Justice Ramakrishnam Raju	BA, BL	Chairman	Non Scientific/ Judiciary/ Legal Expert (Non Institutional)
Dr. B. V. Rama Raju	MS, Mch	Member Secretary	Scientific/ Medicine/ Clinician (Institutional)

#8-2-595/2/B, Care Convergence Centre, Care Foundation, 2nd Floor, Road No. 10, Banjara Hills, Hyderabad - 500 034, Telangana, INDIA
Phone: 91-40-39116069, Fax: 91-39116019, 23355316 | Email: ethicscommittee@carefoundation.org.in

Dr. T. Ravinder Rao	MD	Member	Scientific/ Pharmacology/ Basic Medical Scientist (Non Institutional)
Prof. G. Hara Gopal	PhD	Member	Non Scientific/ Politics/ Social Scientist (Non Institutional)
Dr. Mohan Kanda	M.Sc., PhD, IAS	Member	Lay Person from the Community (Non Institutional)
Sri A. Krishnam Raju	BA, BL	Member	Non Scientific/ Advocate/ Legal Expert (Non Institutional)
Dr. C. Sridevi	DNB, MNAMS, CCDS	Member Secretary	Scientific/ Medicine/ Clinician (Institutional)
Dr. S. Radhakrishna	PhD	Member	Non Scientific/ Statistics (Non Institutional)
Dr. B. Dibbala Rao	MS	Member	Scientific/ Medicine/ Clinician (Non Institutional)
Dr. M. Goverdhan	MD	Member	Scientific/ Medicine/ Clinician (Institutional)
Mrs. Haritha Vijayan	DGN, M.Sc.	Member	Scientific/ Medicine / Nurse Clinician (Institutional)

The Institutional Ethics Committee is working in accordance to ICH-GCP, Schedule-Y, ICMR guidelines and other applicable laws and regulations. None of the study team participating in this study took part in the decision making for the study.

The Ethics committee, after detailed deliberations based on your presentation, review of the documents of thesis protocol and discussion with the IEC full board accorded ethical clearance to conduct the retrospective data collection study at the site Department of Cardiology, Care Hospital, Hyderabad.

Thanking you

Yours faithfully,
For Institutional Ethics Committee

(Dr. B. V. Rama Raju)
IEC Member Secretary

www.ingramcontent.com/pod-product-compliance
Lightning Source LLC
Chambersburg PA
CBHW081747220526
45468CB00008B/2276